Edgar Hamilton Nichols

Elementary and Constructional Geometry

Edgar Hamilton Nichols

Elementary and Constructional Geometry

ISBN/EAN: 9783337215743

Printed in Europe, USA, Canada, Australia, Japan

Cover: Foto ©berggeist007 / pixelio.de

More available books at **www.hansebooks.com**

ELEMENTARY AND
CONSTRUCTIONAL GEOMETRY

ELEMENTARY

AND

CONSTRUCTIONAL GEOMETRY

BY

EDGAR H. NICHOLS, A.B.
OF THE BROWNE AND NICHOLS SCHOOL, CAMBRIDGE, MASS.

NEW YORK
LONGMANS, GREEN, AND CO.
LONDON AND BOMBAY
1896

COPYRIGHT, 1896
BY
LONGMANS, GREEN, AND CO.

TROW DIRECTORY
PRINTING AND BOOKBINDING COMPANY
NEW YORK

PREFACE

THIS book is designed for pupils beginning Geometry at the age of twelve, or even younger. It is based upon the author's class-room experience with young boys during the last twelve years. It is hoped that it may prove an aid to those teachers who do not want a text-book except as a guide to the pupil, to enable him to review the principles developed in the class by answering for himself a set of questions similar to those asked in the class-room, and by working out practical problems illustrative of the principles studied. It cannot be used as a text-book from which lessons are assigned in advance to the class. Each principle should be developed by the teacher and the class, working together—the principle coming at the end, not at the beginning, of the study—and then a lesson from the book, covering essentially the same ground, should be assigned. The book has not been divided into lessons, because sometimes a single paragraph will afford material for several lessons, and, again, several pages can be mastered in one lesson. Each teacher should be free to develop his subject in his own way without artificial lines of division.

The main object of the first year's study is to make the pupils perfectly familiar with the use of their tools, so that in the study of Theoretical Geometry the construction of the figures will present no difficulty. In the pursuit of this object a great many useful working principles are learned, which greatly simplify the more advanced study of the subject.

The proof of principles is introduced very gradually,

and the use made of even the few proofs given must depend upon the age and the ability of the particular class concerned.

From the beginning it is insisted that the pupil shall have a clear idea of the exact direction of any line named with two letters. No slovenliness of thought should be permitted in this matter. For greater accuracy of expression two words have been coined—*symparallel* and *antiparallel*. Parallel lines that have the same direction are called symparallel lines, and those that have exactly opposite directions are called antiparallel lines. Thus, antiparallel lines form an angle of 180° with each other, but symparallel lines form no angle. It is found that this distinction prevents all confusion in the study of parallels, and helps to make the pupils think clearly and express themselves with accuracy. Simple and natural signs are used for these words.

Attention is called to the treatment of areas. The author has found that the subject lends itself especially to the object aimed at in the first year's study. An indefinite field for the ingenuity of the pupils is opened, and often excellent results have been obtained. In his manipulation of areas the ordinary problems of construction (often reserved as the very last subject of Geometry) become a mere matter of course to the pupil, and his familiar acquaintance with valuable principles is assured. A similar treatment of volumes opens a still larger field, if the time allotted to the subject allows it.

A proper use of the Summary will add greatly to the interest and to the value of the subject. In the first place, each teacher can word his definitions to suit his own convictions, and can decide what principles he will take up, and how those principles shall be expressed. In the second place, the pupil feels an added interest in the principles, because he has helped to decide upon the best wording of them. An important feature of the work, therefore, is the training that it gives in accuracy of expression. The teacher ought first to make the pupils express the principles in their own words, and then by interpreting their English literally, to bring out any inaccuracies of expression.

A pupil who has acquired a familiar knowledge of the principles developed in this book should be able to take up the study of Theoretical Geometry, both plane and solid, in a text-book where no complete proofs, but suggestions only, are given to aid in the solution of the more difficult problems and theorems. He should be encouraged to consult, after his own solution, the text-books that contain proofs (of which there should be several kinds in the school library), in order to develop his critical power and to add life to the subject.

The author has in mind the preparation of a Geometry to supplement this book, in which the principles of Plane and Solid Geometry will be developed according to the suggestions just made. He will be grateful to any who will call his attention to typographical errors and ambiguous expressions in this book that may have crept in, in spite of the greatest care on the part of himself and of friends, who have kindly read and re-read the proofs. He is indebted to one of his former pupils for the best of the diagrams.

EDGAR H. NICHOLS.

December, 1895.

CONTENTS

		EXERCISES	PAGE
PREFACE			v
SECTION I.	INTRODUCTORY	1–23	1
SECTION II.	INTRODUCTORY	24–58	5
SECTION III.	LINES	59–74	10
SECTION IV.	ANGLES	75–112	14
SECTION V.	PARALLELS	113–121	22
SECTION VI.	REVIEW	122–139	26
SECTION VII.	PARALLELS (Continued)	140–171	28
SECTION VIII.	TRIANGLES	172–195	32
SECTION IX.	REVIEW	196–209	39
SECTION X.	ISOSCELES TRIANGLES	210–246	41
SECTION XI.	TRIANGLES (Continued)	247–280	48
SECTION XII.	QUADRILATERALS	281–323	55
SECTION XIII.	DIVISION OF LINES	324–366	65
SECTION XIV.	MULTIPLICATION OF LINES; AREAS	367–398	72
SECTION XV.	MOULDING AREAS; EQUIVALENT FIGURES	399–436	79
SECTION XVI.	PECULIARITIES OF SQUARES	437–532	87
SECTION XVII.	AREAS OF SIMILAR FIGURES	533–565	104
SECTION XVIII.	CIRCLES AND INSCRIBED ANGLES	566–625	111
SECTION XIX.	VOLUMES	626–678	123
INDEX			137
SUMMARY OF FACTS, DEFINITIONS, AND PRINCIPLES			139

ELEMENTARY

AND

CONSTRUCTIONAL GEOMETRY

SECTION I.

1. In the group of drawings on the third page are outlines of several familiar bodies or **solids**. Learn their names, and find as many examples as you can of each kind of solid in the objects about you. Can you name any other solids not represented there?

2. One of the aims that we shall keep before us in our first study of Geometry will be to discover the characteristics of these and other common solids. We shall study at first with special care the nature of their bounding surfaces and of their bounding lines; we shall learn how to make accurate drawings and measurements of them; and we shall study the peculiarities of form in the lines, surfaces, and solids.

3. In looking at the physical solids about us, we think not only of their form, but also of the material of which they are made, of their color, and of their fineness of finish; but in dealing with geometrical solids, we think of the form only. If, for example, you move a cube of glass from its place, and then imagine the space which it first occupied to be bounded by surfaces and lines, it will help

you to form an idea of a geometrical solid. You must, however, be on your guard against one danger. Your glass cube, however skilfully made, will have imperfections in it; the lines will not be true, or the surfaces will not be perfectly smooth: but the geometrical cube of your imagination is absolutely perfect; every line is true, every surface perfectly smooth. It is a great aid to the imagination to use physical solids for illustration; but you must never forget that the geometrical solids which they represent, and which you are to study, have none of their imperfections. In all your drawings, too, you must aim to have your work as near perfection as is possible; to this end you must provide yourselves with well-sharpened hard pencils, a good ruler, and a strong pair of compasses. It is better to have one pencil set aside for drawing only, and to have that sharpened with a chisel edge for use with a ruler. But you must never imagine that the solids have any of the faults of your drawings.

4. The **geometrical solid**, then, is a portion of space bounded by imaginary surfaces. It is limited in three directions at right angles to each other; or, in other words, it is said to have three dimensions, length, breadth, and thickness. Sometimes height or depth is used in place of thickness. Can you think of any practical examples where height or depth would be a better word than thickness? Do you know of any word to take the place of breadth?

5. Examine a cube (Fig. 1). Which dimension do you call its length, which its breadth, and which its thickness? Is there any difference in the magnitude of these dimensions? Can a cube have unequal dimensions?

6. Can you draw a cone or a cylinder with equal dimensions?

7. What do you consider to be the height of a pyramid?

8. Can you suggest any practical way of finding the height of a (physical) pyramid?

9. Which of the solids represented *must* have equal dimensions from their very nature?

CONSTRUCTIONAL GEOMETRY

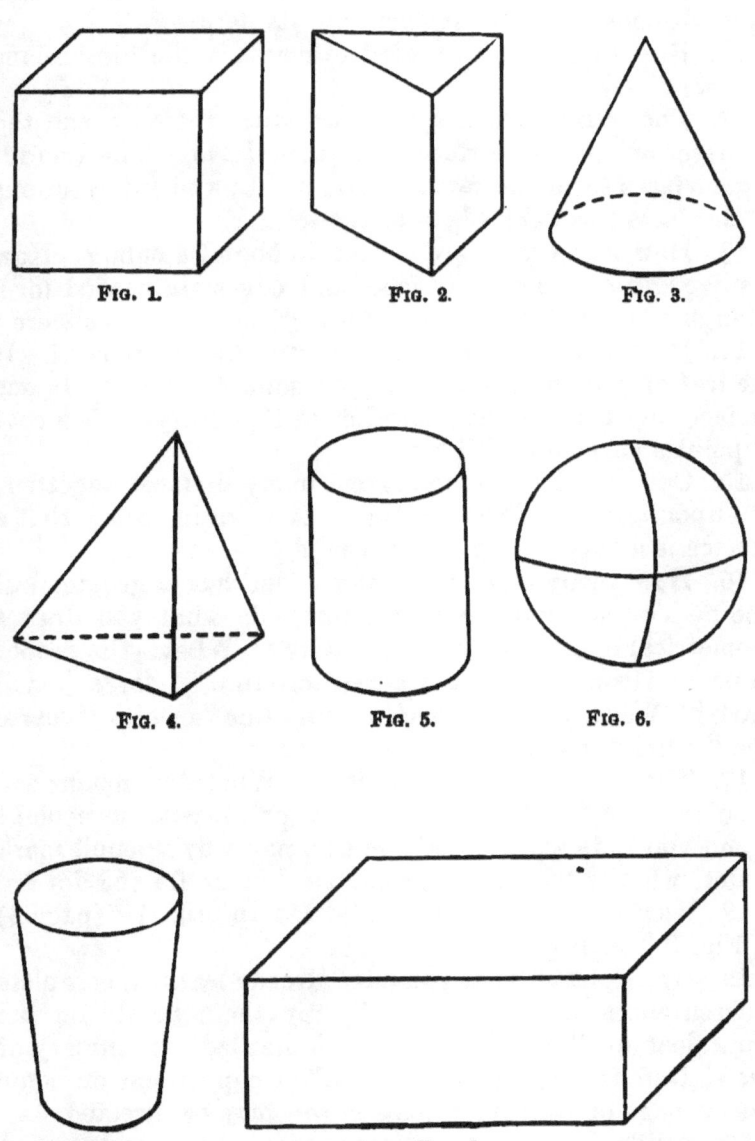

FIG. 1.   FIG. 2.   FIG. 3.

FIG. 4.   FIG. 5.   FIG. 6.

FIG. 7.   FIG. 8.

10. Is there any solid in the group which cannot have equal dimensions without changing its nature?

11. How would you proceed to measure the dimensions of a cone of wood?

12. The boundaries of a solid are called **surfaces**, and the intersections of the surfaces are called **lines**. The bounding surfaces form the **faces** of the solid; the intersections of the faces form the **edges** of the solid.

13. How many faces are needed to bound a cube? How many edges? How many faces and edges are needed for a triangular prism? for a cone? for a cylinder? for a sphere?

14. How many and what dimensions has a surface? Is the leaf of your book a surface or a solid? Why? Is one surface any thinner than another in Geometry? Is a coat of paint a surface? Why?

15. Can you, by placing a great many surfaces together, one upon another, form a solid? Is it right to say that a surface is a very thin piece of a solid?

16. How many and what dimensions has a geometrical line? You speak of drawing a line. Is what you draw a geometrical line? What is it? Why? What is the proper name in Geometry for a "telegraph line"? for a pencil mark? When you speak of a "fine line" or of a "coarse line," what are you really describing?

17. The ends of a line are **points**. What dimensions can a point have? Can you make a point with a pencil? When you indicate the position of a point by a pencil mark or dot, what is the proper geometrical name for the dot?

18. How many points are indicated in Fig. 1? (page 3) in Fig. 3? in Fig. 4?

19. By means of what you have already learned, complete the sentences that follow, and copy them neatly on the blank leaf at the end of the book headed "Summary of Facts, Definitions, and Principles," or copy them on some special page of your note-book as you may be directed.

20. Solids have........dimensions, and are bounded by........

21. Surfaces have........dimensions, and are bounded by........

22. Lines have........dimensions, and are bounded by........

23. Points have........dimensions; they show position only.

SECTION II.

24. If it is not possible to make points, or to draw lines, are we not blocked at the very beginning of our course? Not so, if we remember what our drawings really amount to; they are merely aids to the imagination; the dot, which stands for a point, enables us to be sure that all are thinking of the same point or position; the line A B is an aid to the eye, and makes it easier to remember that a line is imagined between the points A and B; but the imagination should always go behind the poor imitation of a line to the absolutely perfect line represented by it. If the line is not needed to aid the imagination, it is better not to draw it; for instance, in the figures of Section I. a great many lines are left for the imagination to supply, because the mind does not need them to form a true picture. Give examples of omitted lines in those figures. With the explanation just given we can speak of drawing lines and of making points without causing any confusion.

25. Imagine a point A to move from its position to the position B. It is clear that you can imagine it to move along several paths. How many? How many dimensions will any one of these paths have? Why? What, then, is the proper name for the imaginary path? See 20-23.

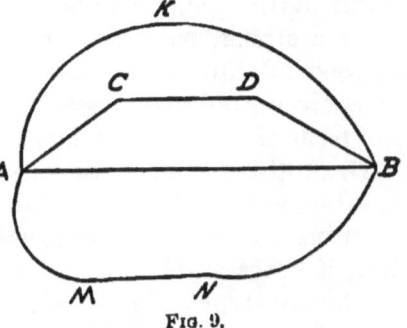

Fig. 9.

26. We can, therefore, describe a line in another way, which is better for our purposes upon the whole, as follows: **A line is the path traced by a point as it moves from one position to another.** Record this definition in the "Summary."

27. There are four paths or lines between A and B pictured in the figure; the upper path is called a **curved line**; the next, A C D B, is called a **broken line**; the next, A B, a **straight line**; and the bottom one a **mixed line**.

28. Write out a definition of each kind in your own language, aiming to describe the characteristics of each so clearly that there will be no confusion among the kinds. Bring your results into the class-room, and after deciding upon the best wording, learn the definitions which are selected as the best. After this, if a line between two points is mentioned, a straight line is always to be imagined unless otherwise specified.

29. Select examples of the different kinds of lines from the figures 1 to 8 on page 3.

30. Can you, without removing your pencil from the paper, and without going over the same path twice, trace the outline of Figs. 1, 2, 3, 5, and 6?

31. Can you trace all four paths in Fig. 9 with a continuous movement?

32. With your compasses trace a **circumference.** Write as good a definition as you can of a circumference viewed as the path of a moving point.

Draw three concentric circumferences. (Look up the meaning of the word concentric, if necessary.)

33. Construct Fig. 10 with your compasses, beginning

Fig. 10.

with the central circumference and using one **radius** for all the circumferences. What is meant by the **radius** of a circumference? Take great pains with this figure, and do not give up until you have an exact and neat copy of it.

34. Construct Fig. 11, beginning with three points one inch from each other.

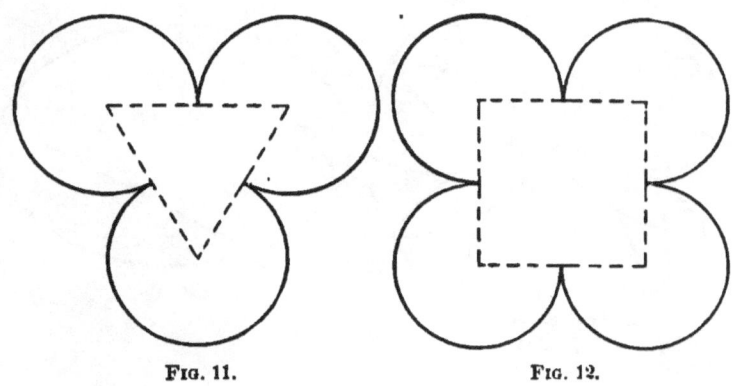

Fig. 11. Fig. 12.

35. Construct Fig. 12, putting the centres at the corners of a square an inch each way.

36. Construct a spiral (Fig. 13) by using O as the centre for the first semi-circumference A C B and every alternate one, and by using A as the centre for the second semi-circumference B D E and every alternate one.

37. Construct Fig. 14.

38. Construct Fig. 15, by first placing two drawing-pins or small tacks about 2½ in. apart at M and N; then, fastening a piece of fine

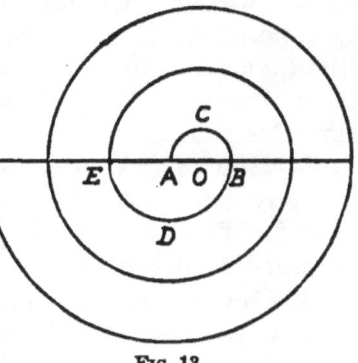

Fig. 13.

thread 3 in. long to the two pins, with your pencil pressed firmly against the thread, trace the curve. This curve is called an **ellipse**. Try the experiment of putting the pins nearer together without altering the length of the string.

Put the pins close together; what kind of a curve do you now trace? The ellipse, in addition to being a beautiful curve, has a peculiar interest to us because it is the path

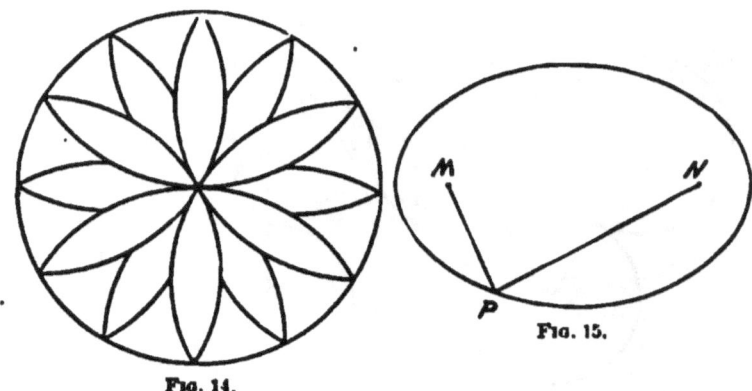

Fig. 14.   Fig. 15.

which the earth traces as it moves about the sun. The sun is at one of the places represented by the pins.

39. Imagine a point to move from A to D (Fig. 16), tracing the line A D, and then imagine the line A D to move, without tipping, to a new position, B C. Can the line take more than one path? How many? Make illustrations for your answers.

40. How many dimensions has any one of these paths? Why? What should the path be called? See 20–23.

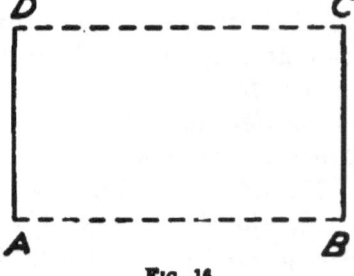

Fig. 16.

41. Can you move a line in such a way that it will not generate a surface?

42. Can you trace the lateral, or side, surfaces of Fig. 1 by the movement of a line as described in 40? of Fig. 2? of Fig. 4? of Fig. 5?

43. How would you move a line to generate the lateral surface of Fig. 3?

44. What kind of line would you move to generate the surface of Fig. 6, and in what way would you move it?

45. Can you generate the lateral surface of Fig. 5 by the movement of a curved line?

46. Draw a curved line A B C D (Fig. 17). Move a line A P, without tipping it, so that A shall remain on the curve. Will a surface or a solid be generated? Why?

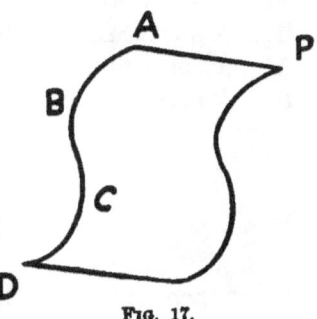

Fig. 17.

47. Compare the surfaces which have been generated in exercises 39–42, and tell which are plane and which are curved surfaces. Write a good definition of **a plane surface** that will distinguish it from a **curved surface**. Select the best definition from those handed in by the class, and record it in the "Summary."

Suggestion: Think how a carpenter tests with his straight-edge the evenness of a floor or table.

48. Imagine the point A to move to D (Fig. 18), generating the line A D; the line A D to move to B C, generating the surface A B C D; and the surface A B C D to move to the position E F G H. Could the surface take more than one path in moving to its new position? Illustrate your answer. How many dimensions in any one of the paths?

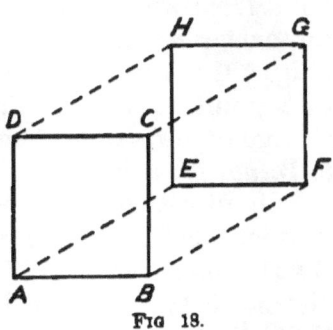

Fig. 18.

49. Could you generate by the motion of a surface the solids of Figs. 1, 5, 6, and 8?

50. In what way could you move a surface to generate the solids of Figs. 3, 6, and 7?

51. In what two ways could you generate the cylinder, Fig. 5?

52. Can you move a surface so that it will not generate a solid ?

53. Review from 24, and complete the sentences that follow, following the directions given in 19 :

54. A line is the path traced......

55. A straight line is the path traced......

56. A surface is the path traced......

57. A solid is generated when......

58. A solid of revolution is generated when......

SECTION III. LINES.

59. Since a straight line is merely the path traced by a point as it moves, without change of direction, from one position to another, the line can have only two characteristics, which are determined by the direction in which the point moves and by the distance which it moves. It is very important in naming a line to indicate, if possible, the direction of it ; it is, therefore, important to put first the point from which the moving point is imagined to start ; for example, the line A B (Fig. 19) is the path traced by a point moving, without change of direction, *from A to* B, *not from* B *to* A.

The direction of the line A B is, therefore, *from A to* B ; the length of A B is the distance between A and B. Describe the lines M N, M A, N M, B M (Fig. 19) in a similar way. It ought not to be necessary for you to represent the lines in this case to get a clear idea of them. The points that limit the lines should enable you to form a distinct picture of the lines in your mind.

60. In what ways does the line B A differ from B M ?

How does M N differ from M N P? (Fig. 19). How does M N differ from N M?

61. How many different straight lines can you draw *from* M *to* N? (Fig. 19). Can you give a reason for your answer?

An **axiom is a truth that is assumed as self-evident**, so that it needs no reason. Can you state the axiom that you have made use of in your answer to 61?

62. Can you draw *through* M and N more than one straight line? (The straight lines that you draw must differ in some way.) Can you draw *between* M and N as the limiting points more than one straight line? Compare line M N P with M N, and N M with M N before giving your answer, and be sure that the axiom of 61 is worded exactly.

63. Sometimes, when the direction of a line is of no consequence, a line may be named by a single letter; for example, A B (Fig. 19) might be called "*l*." (Small letters are used when it is desired to name a line in this way.) It is clear, however, that the name "*l*" gives no suggestion as to whether the line points *from* A *to* B, or *from* B *to* A; that is, there is doubt about one of the two characteristics that are possible for a line. Therefore, unless the direction is of no consequence, or unless the figure in itself makes doubt impossible, it is unwise to use a single letter in naming a line. Of course a line should never be named B A, if A B is the line in one's mind. Beginners cannot be too careful about this point.

64. On one indefinite straight line mark two lines that shall differ in length, but not in direction; also on one straight line mark two lines that shall differ in direction, but not in length.

65. On one straight line mark two lines that differ both in length and in direction.

66. In the exercises that follow pay no attention to the direction of the lines, but try to estimate accurately the lengths of the lines. Estimate the shorter lines in inches or centimeters, the longer lines in feet or decimeters.

Make your estimate first, and record it; then measure the line, and record the result in a second column; in a third column put the error made.

Make five lines, *a, b, c, d, e*, of different lengths and in different positions; estimate their lengths in inches, and record results as explained above.

67. Make a line, and try to make five others, in various positions, of the same length; record your errors in millimeters.

68. Make a six-sided figure with sides varying considerably in length. Estimate the lengths in centimeters, and record estimated lengths, true lengths, and errors in three columns as before.

69. Let one of the class put ten different lines on the board, keeping to himself the true lengths of the lines; let the rest of the class hand in to him their estimates in feet, and let him average the errors made by each one.

70. Make a horizontal line; try to make a vertical line and a slanting or **oblique** line of the same length as the horizontal line. Record errors in millimeters. It is best to measure a line with your compasses, and to apply your compasses to the ruler.

71. Estimate in inches the height of a chair (to the seat), the dimensions of a window-pane, the width of a door. Record results as before.

NOTE.—Exercises of a similar nature can be given profitably, until each pupil becomes fairly expert in estimating lengths.

72. In comparing the lengths of two lines, *a* and *b*, we say that *a* is equal to *b*, *a* is greater than *b*, or *a* is less than *b*, according to the facts, recording the result in one of the following ways: $a = b$, $a > b$ or $a < b$. Note the meaning of the signs, which are used in comparing any two magnitudes.

Make two broken lines of three parts each; measure the total length of each line, and put the proper sign of the three signs just explained between the two results.

73. Make three different lines a, b, and c; make one line equal to $a + b + c$. Make another line equal to $a + b - c$; another equal to $a - b + c$. Put the proper sign between $a + b - c$ and $a - b + c$.

74. Making use of the lines a, b, c, d, and e in Fig. 20, find the line equal to $a - b + c - d + e$; find also $b - 2c$

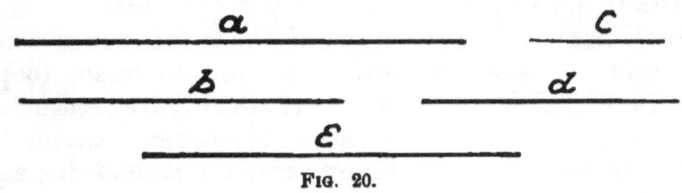

Fig. 20.

$+ d - 3d$; $a - 3c + b - e$. Record the length of the answer in millimeters for each case. Since all of the class are using the same lines, the answers should come out alike. Do they? If not, can you account satisfactorily for the differences, supposing each one to have taken the greatest possible care?

Although measuring a line with compasses and ruler seems a very simple thing in theory, in practice it is a very hard thing to do, when great accuracy is required. This is due partly to the fact that the points, which represent the ends of the lines, are very far from being true points, and partly to the inaccuracy of our tools. Two lines that seem to be of the same length, when measured with an ordinary ruler, might prove to differ distinctly when measured with a more delicate tool such as the micrometer screw. Do you know what a micrometer screw is? If not, let some of the class find out what it is, for the next lesson, securing one, if possible, for actual use. Do you know of any other device for measuring lengths with great accuracy?

The result of your work in the exercises of 74 and what has just been said should teach you that it is not safe to rely wholly upon measurement when it is to be decided whether two lines are of equal length or not. Lines that seem to the eye or to the tools to be equal are often unequal; while lines that seem to be unequal are often equal. The sides of a true square, for instance, we know must be equal from the nature of the square, but the sides of the figure that we draw to represent the square may be unequal. You must, of course, use your tools to draw figures to aid your imagination; but you must think more of the imaginary figures than of the actual figure seen by you, and you must not think it a sufficient reason to prove two lines equal in length to say that you have measured the lines and found them to be equal. Some reason, based upon the nature of the figure as in the case of the square above, must be given, if possible.

SECTION IV. ANGLES.

75. In the last few exercises you have thought only of the length of the lines; now consider their direction. If two lines do not point exactly alike, they are said to make an **angle** with each other. Thus in the oblong A B C D E F G (Fig. 21), the edge A B is said to make an angle with the edges A D, B C, E F, B E, and D G, because it does not point as they do.

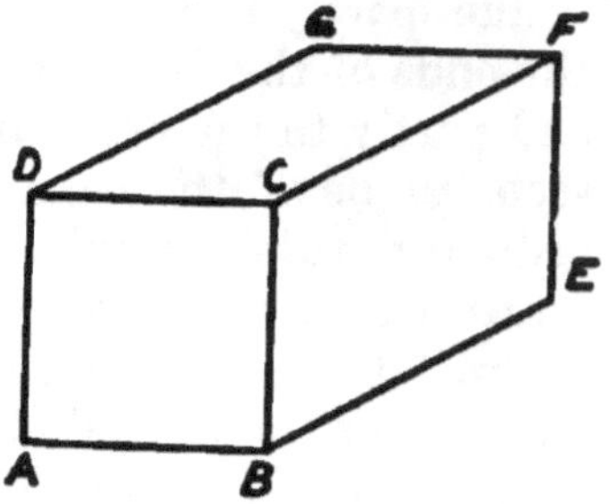

Does A B make an angle with D C? Why?

Does A B make an angle with G F? with C B? Why?

You have very likely used the word "angle" in the place of the word "corner;" thus, in saying that a house was in the angle of two streets, you meant that it was just at the corner in the space between the two streets; or, perhaps, you have thought of an angle as the *space* included between two lines. But in Geometry

Fig. 21.

the angle has to do solely with the direction of the lines; if the lines point alike, they form no angle; if they differ in direction, they form an angle. To make an angle larger, we do not produce its sides that they may inclose more space between them, but we make the difference in direction greater by turning one of the lines away from the other. Starting with your compasses closed, turn one leg gradually away from the other: the two legs at once begin to point in different directions so that they form an angle with each other; as you keep on turning, the legs point farther and farther away from each other, going through the various positions shown in Fig. 22, and the angle grows larger and larger. In time the turning leg points exactly opposite to O A in the direction O F. If we turn the leg still farther to the position of O G, there is not so much difference in the way the legs point as there was when the turning leg was at O F; in other words, the angle is growing smaller again; if we keep on turning, presently the leg will point again as O A does, and there will be no angle formed.

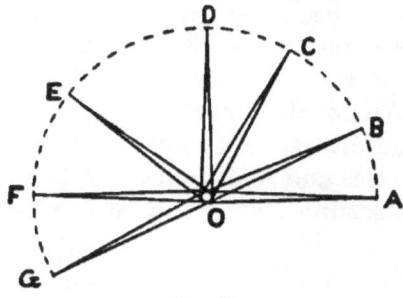

Fig. 22.

76. Enough has now been said *about* an angle to enable us to say what the angle *really is*. **An angle is the difference in direction of two lines.** It indicates how much you must turn one of the lines to make it point like the other. The amount that one of the lines has to be turned to make it point like the other is a very good measure of the size of the angle. The angle is not the corner or the place where two lines meet; it is not the space between the lines; it is not anything that you can see: but it is a difference to be determined by the judgment, as any other difference is determined. Perhaps the best illustration from familiar things, to help us to form a clear idea of what

an angle really is, is to be found in colors or in sounds. We see two shades of red, for instance. We see that there *is* a difference, but we do not see the difference itself; that is something to be determined by judgment. If we are well trained in distinguishing colors, we can estimate the difference in shade with great accuracy. Again, when we hear two different notes on the piano, we can estimate the difference in pitch exactly, if our ears have been trained to distinguish sounds. In like manner, if we train ourselves to distinguish differences in direction, we can estimate the size of the angle with surprising accuracy. We must, however, have some means of expressing our estimates. In colors we estimate by shades; in sounds, by tones and semi-tones. What is to take the place of these measures in the case of angles?

77. In your experiment in 75 (Fig. 22), as you turned the leg of the dividers, you made O B point more and more away from the direction of O A, but nearer and nearer to the direction of O F, which has the direction of A O exactly opposite to that of O A. There must have been one position of O B when it diverged from O A just as much as it did from O F. In that position (O D in Fig. 22) it made a **right angle** with O A and also with O F, so that the adjacent angles F O D and A O D were equal. In Geometry the **right angle**, or the angle formed when one line meets another so as to make the adjacent angles equal, serves to divide angles into two large classes. Any angle less than a right angle is called an **acute** angle; any angle larger than a right angle is called an **obtuse** angle. It is also common to call all angles that are not right angles, whether acute or obtuse, **oblique** angles.

78. Select as many right angles as you can in Fig. 21. In an ordinary room the two walls intersect in a corner line which makes a right angle with the edges of the floor. Imagine several lines on the floor starting from a corner. What angles do these lines make with the corner line of the two walls? Can you draw on the floor or ceiling a straight

line that does not make a right angle with this corner line?

The right angle is perhaps the most interesting of all the angles. It is the angle that the carpenter must understand thoroughly; if he makes mistakes in his right angles, his work can never look well. How many devices of the carpenter for getting true right angles do you know?

79. Another angle that plays an important part in Geometry is the angle formed when one line points in a direction exactly opposite to that of another line. This angle is called a **straight** angle; it is the largest angle dealt with in elementary Geometry; it indicates that one of the lines would have to make a half revolution to point like the other. Select examples of straight angles in the room.

80. The division in 77 of angles into two large classes is not definite enough for our purposes. For greater definiteness in estimating angles, a semi-circumference is divided into one hundred and eighty equal parts called degrees. The number of degrees that O A (Fig. 23) would have to pass over in reaching the position of O C

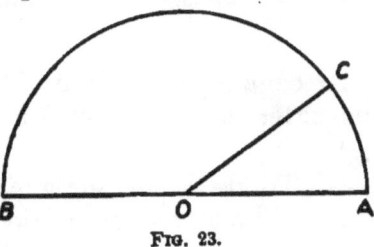

Fig. 23.

is an exact measure of the size of the angle between O A and O C. If it is found that O A would have to turn past thirty divisions, the angle is said to be an angle of thirty degrees— written 30°. And O A would make an angle of 30° with any line pointing exactly like O C, no matter where in space it might be.

81. How many degrees in every right angle? How many degrees in the angle between O A and O B? (Fig. 23). How many degrees must there be in any angle to make it obtuse? How many degrees in a straight angle?

82. What do you mean by saying that the angle between O A and O D is 70°? 60°? 90°? 180°? Answer each part in a full, clear sentence.

83. The instrument used for measuring angles is a graded half circle of metal or cardboard called a **protractor** (Fig. 24). The centre is clearly marked by a notch. The

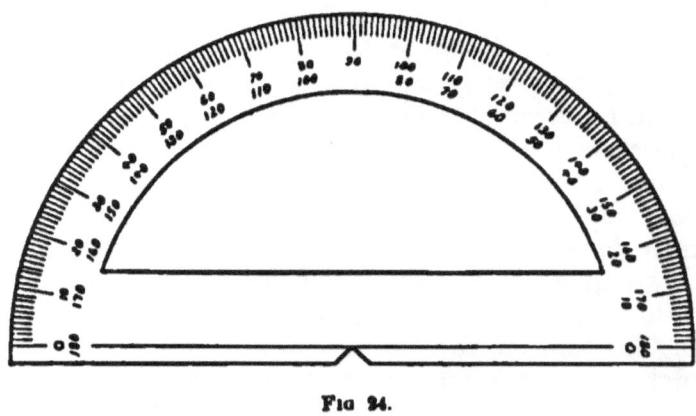

Fig. 24.

inner edge is graded into 180 equal parts from right to left; the outer circle is graded into 180 equal parts from left to right.

84. To make an angle of 70° with your protractor first draw a line of any convenient length (M N Fig. 25), then place your protractor with the centre mark at M, and with one of the zero lines along M N; next, put a dot P at the point exactly opposite the line marked 70° on your protractor, remove your protractor, and join M with P. If you wish the line to be above M N, you should use the 70° mark of the inner circle; if you wish it to be below M N, use the outer circle. Draw both cases.

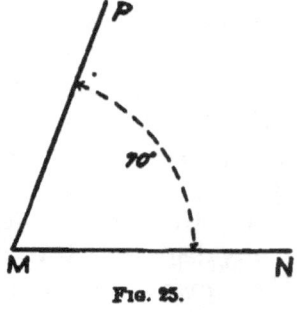

Fig. 25.

85. With your protractor make angles of 22°, 60°, 45°, 125°, and 170°.

86. Draw two lines, A B and A C crossing each other at A

(Fig. 26). Explain fully, step by step, the way in which you would use your protractor to measure the angle between A B and A C. Measure it twice, placing the zero line of your protractor first on A B and then on A C.

87. For the sake of brevity the following signs are used: ∠ = angle; ∠s = angles; ∠ C A B or ∠ B A C = angle between A B and A C. In using the last sign be careful to put the letter naming the point at which the two lines meet *between* the other two. The angle between A B and A C could not be written ∠ A B C or ∠ A C B; A, the meeting-point of the two lines, called the **vertex** of the angle, *must be* the middle letter. What is the plural of vertex? If only one angle is formed at a vertex, the angle may be named by the vertex letter. In explaining the meaning of ∠ B A C, you must remember to read the line *from* the vertex *out;* for instance, ∠ B A C is the difference in direction between A B and A C; it shows how much you must turn A B to make it point like A C. It would not do to substitute B A for A B, or C A for A C. If you think for a moment how much you would have to turn A B to make it point like C A, you will see the need of being careful.

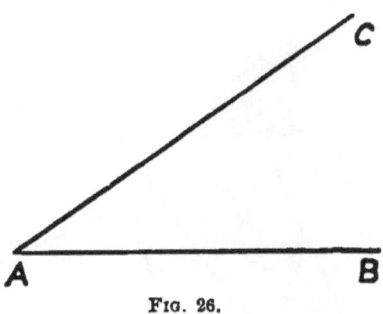

Fig. 26.

88. Explain the meaning of ∠ M N R, ∠ N M S, ∠ P T Q, without drawing the lines that form the angles.

89. Make any figure with five sides. Estimate the angles of the figure; record your estimate in one column; the true value, found with your protractor, in another; and the error in a third column.

90. Repeat 89, with a four-sided figure.

91. How would you measure the angle between A B and C A (Fig. 26)?

First estimate it and then measure it.

92. Estimate the value of the angles 1, 2, 3, 4, 5, 6, and 7 in Fig. 27, recording estimates, true values, and errors as in 89.

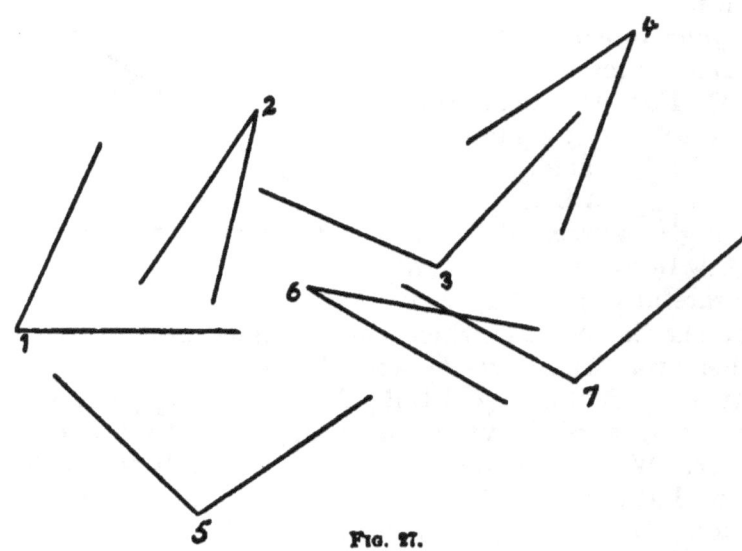

Fig. 27.

93. Make angles of 60°, 20°, 110°, 45°, and 90° free hand. Test your accuracy with the protractor. Repeat this exercise, drawing lines that do not meet.

94. Divide an angle into two equal parts with your protractor.

95. Divide an angle into three equal parts with your protractor.

96. Make an angle C A B (Fig. 28) with the aid of your protractor, and a line O R. Make an angle with vertex at O equal to ∠ C A B.

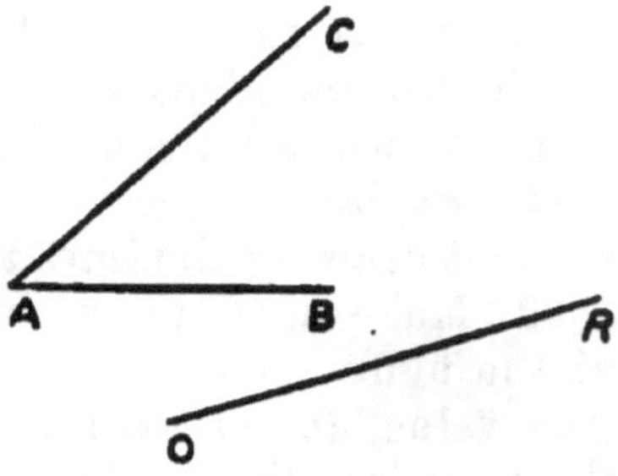

Fig. 28.

97. Make a five-sided figure or **pentagon**. Make a second pentagon with the same angles but different sides.

98. Make two angles that to your eye seem to be of the same size. Can you test their equality by using your compasses without the aid of your protractor? Explain your steps carefully.

99. Can you make one angle equal to another without your protractor? Give three illustrations.

100. Make a four-sided figure, or **quadrilateral**, and make another quadrilateral with different sides but the same angles, using your compasses and ruler only.

101. Make two angles, A B C and D E F; make an angle that you think is equal to the sum of $\angle$ A B C and $\angle$ D E F. Test your accuracy with the help of your compasses.

102. Can you explain how to double an angle A B C?

103. Can you add two angles? three angles? Make your work clear in all the steps.

104. Draw two angles; add them with the aid of compasses and ruler only; test your result by finding the number of degrees in each angle with your protractor.

105. Make a figure with three sides, called a **triangle**; add the three angles formed at the three corners. How many degrees are there in the angle which you get for your answer?

106. Make four triangles of different sizes and shapes. Add together the angles of each one, and note the number of degrees in the resulting angle each time. In adding the angles use ruler and compasses only. In getting the number of degrees use a protractor, if necessary. Do your experiments seem to point to any principle?

107. Cut out a paper triangle A B C; cut off the corners along wavy lines, and place the vertices A, B, and C together. Does this experiment confirm the previous experiments?

108. Make a quadrilateral, and add the angles formed. Try two figures differing in shape and size. If one line should turn through all the angles of the quadrilateral, how far would it turn? Measure each angle with a protractor: find the number of degrees in all four angles of each figure.

109. Make a five-sided figure, and add the angles formed.

110. Can you draw any conclusion about the sum of the angles of a figure of many sides?

111. Verify the results of 109 and 110, by cutting off the corners of paper figures of the corresponding number of sides. Be careful to mark the corners before cutting.

112. Can you describe a way of finding the difference between two angles with compasses and ruler only? Make two angles, and find an angle equal to their difference.

SECTION V. PARALLELS.

113. We have seen that, if two lines point in the same direction, they cannot form an angle. Do you remember why? We have seen also that, if they point in exactly opposite directions, A B and D C, for example, in Fig. 29, they form a straight angle (79).

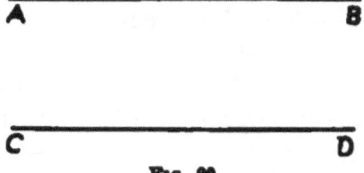

Fig. 29.

Whether two lines point the same way or exactly opposite ways, they are said to be **parallel**, because they run *by the side of each other*. (The name is taken from two Greek words.) It is a great convenience to be able to distinguish the two cases. When two lines are parallel and point the same way the following sign, ↑↑ , will be used, and the lines will be called **symparallel** lines. For instance, A B ↑↑ C D is to be read A B points in the same direction as C D, or, more briefly, A B is symparallel to C D. When the two lines are parallel and point in opposite directions, the sign ↑↓ will be used, and the lines will be called **antiparallel** lines. Thus A B ↑↓ D C means that A B runs in the opposite direction to D C, or A B is antiparallel to D C. If single letters are used, the direction of the lines is not indicated; so that the sign without the arrow-heads can be used. Thus $a \parallel b$ mean that the line "a" points either in the same direction as "b" or in the opposite direction to "b."

114. Draw a line A B, and choose a point C anywhere outside of A B (Fig. 30). Can you draw through C two distinct lines diverging from each other and yet symparallel to A B? It does not take long to see that the problem is impossible, but when we try to give a reason for it, we find that it is one of the truths that is self-evident or axiomatic. It is equally self-evident that *one* line can be drawn through C in the plane of the paper pointing like A B. The axiom can be stated as follows: **Through any point in space one line, and only one line, can be drawn in any desired direction.**

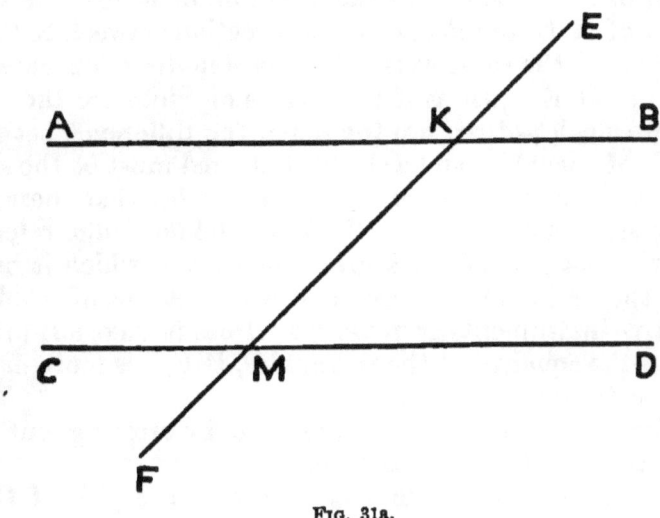

FIG. 31a.

115. Imagine two lines, A B and C D, to have the same direction (Fig. 31a), and imagine that E F crosses these lines at K and M respectively. At once we see the possibility of a great many angles. There are six angles with their vertices at K, and six with their vertices at M. Can you account for them all?

Can you see, without the aid of your protractor or of your compasses, that some of the angles *must be* equal from the nature of the case?

Look at the angles E K B and E M D, for instance; $\angle$ E K B shows you how much you must turn K B to make it point like K E; $\angle$ E M D shows you how much you must turn M D to make it point like M E. Now the lines K B and M D point in the same way at the start; they also point in the same way when in their final positions, K E and M E. Therefore the amount of turning has been the same in both cases, or, in other words, $\angle$ E K B = $\angle$ E M D. Another way of looking at the same thing is this: since the direction of K B is the same as that of M D, and since the direction of K E is the same as that of M E, the *difference* in direction between K B and K E must be the same as the *difference* in direction between M D and M E; just as if two shades of white are the same and two shades of red are the same, the difference between a shade of the white and a shade of the red must be the same in one case as in the other. Our knowledge that these two angles are equal belongs to the kind of knowledge referred to in the last part of 74, knowledge, that is, which is based upon the nature of the case, and which is not affected by defective instruments or drawings. Does it make any difference in the equality of the two angles, if F E is more or less inclined to the parallel lines?

Make the following principle true by crossing out the words that need to be crossed out:

Two angles are or are not necessarily equal, if their sides have the same directions respectively.

Learn the principle by heart, and also copy it in the "Summary."

116. Write the same principle almost entirely by signs, which have already been explained.

117. Select other pairs of angles in Fig. 31a to which the principle applies.

118. What can you say about the angle A K B? What about the angle E K M? What about their relation to each other? Does your knowledge of the facts depend upon measurements, or upon the nature of the figure? Can you put what you have learned in this exercise into the form of a principle? Select for the "Summary" the wording finally agreed by the class to be the best.

119. ∠ B K M is one part of a straight angle. What is the other part? Two answers can be given to this question. Can the answers have different values? Why? When two angles together make a straight angle, the angles are said to be **supplements** of each other. Can one angle be the supplement of two unequal angles? Why? What is the supplement of 75°? of 110°? of 82°? of 60°? Draw the supplement of each angle of Fig. 27.

120. There is an axiom that **if equals are taken from equals, the remainders are equals.** Do you understand what it means, and do you see that the truth has anything to do with exercise 119? with the principle of 115?

121. Select in Fig. 31a as many pairs of supplementary angles as you can. Note specially those which are the supplements of the same angle. What can you say about the direction of their sides?

Your study of the last three articles should enable you to make the following statement true: **Two angles are or are not necessarily equal if their sides have opposite directions respectively.** Record the principle in the "Summary." Select as many pairs of equal angles in Fig. 31a as you can that are equal by application of this new principle.

If A K M, Fig. 31a, is 25°, find all the other angles in the figure.

If L A K M = $\frac{1}{2}$ L M K B, find the number of degrees in all the angles.

If L B K M = 4 × L C M F, find the number of degrees in all the angles.

Scholium.[1] The angles that are formed when one line crosses two other lines are of such great importance in Geometry that they have received special names. Those without the two lines, or 1, 2, 7, 8 in Fig. 31b, are called **exterior** angles; those within the two lines, or 3, 4, 5, 6, are called **interior** angles; 2 and 8 or 1 and 7 are called **alternate-exterior** angles; 3 and 5 or 4 and 6 are called **alternate-interior** angles; while 1 and 5, 2 and 6, 3 and 7, or 4 and 8, are called **corresponding** angles.

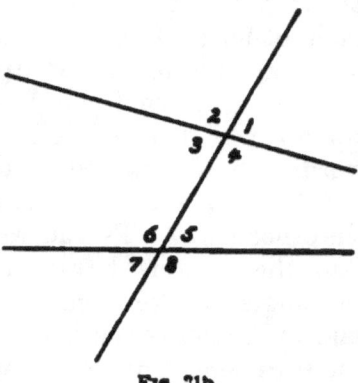

Fig. 31b.

SECTION VI. REVIEW.

122. What is the difference between a geometrical solid and a physical solid?

123. What is a plane surface? How is a plane surface tested practically?

124. How many kinds of lines can you draw between two points A and B? Which is the shortest of them all? How do you know that it is the shortest? How does the straight line A B point—from B to A? or from A to B?

125. Give definitions of concentric and adjacent, with illustrations of their use in Geometry.

126. What is an axiom? Give any axioms that you have learned.

127. Explain how you use a protractor.

128. Write in sign language: A B is longer than C D.

---

[1] A scholium is a remark containing some explanation.

The angle A B C is less than the angle R S T. A B points in the same direction as C D.

129. Write in good English the following:

∠ A B C = ∠ E F G because B A ↑↑ F E and B C ↑↑ F G, or because their sides ↑↑ respectively.

130. What is a pentagon ? a quadrilateral ? a triangle ? What is a straight angle ? an obtuse angle ?

131. Does a straight line change its direction in any part of its course ? Can you place two points so that you cannot draw a straight line through them ?

If two straight lines pass through the same two points can they diverge from each other ? See 113.

132. Can you think of a way of testing the straightness of your ruler's edge by two points on your paper ?

133. What is an angle ? Can you really draw an angle ? What is understood by "drawing an angle" ? What difference in the size of the angle does the length of the sides make ?

134. Place five points so that no three shall be on the same straight line. How many straight lines can you draw that shall each contain two of these points for its ends ? Can you place the five points so that only four lines can be drawn each containing two, and only two, points of the five as the limiting points ?

135. Make as clear as you can the difference between the straight angle A B C and the straight line A B C.

136. What is the vertex of the angle M P N, and what are its sides ? When is ∠ M P N a right angle ?

137. When are two angles supplements of each other ? Give numerical examples and geometrical examples.

138. In how many points can two straight lines cross each other ? two circumferences ? In how many points can a straight line cross a circumference ?

139. In what cases can you say that two angles are equal, without measurement with protractor or compasses ? Write out the principles of 115 and 121, making use of the terms explained in the scholium to 121.

SECTION VII. PARALLELS CONTINUED.

140. Draw two lines A B and A C (Fig. 32); choose any point P on A C. How many lines can there be radiating from P that will make the same angle with P C that A B does with A C? Answer first with the understanding that the line can be drawn from P in any direction in space; then answer the same question with the understanding that the line must be drawn on the same plane surface with A B and A C, for instance, on the plane of your paper. Actually draw the lines in the second case.

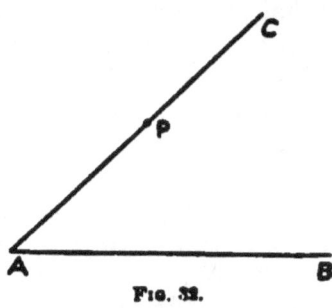

Fig. 32.

141. Draw two lines, A B and A C, as in Fig. 32. At the point P, with the aid of your protractor, make an angle equal to ∠ C A B, drawing your second line P R so that it is on the same side of A C with A B. Since P R will have to diverge from P C just as much as A B does, and since P R and A B are on the same side of A C and in the same plane, they must point in the same direction, or, in sign language, if ∠ C P R = ∠ C A B, P R ↑↑ A B.

142. Could two lines possibly point in the same direction and not be in the same plane with each other? See 114.

143. Repeat the work of 141, using your compasses instead of your protractor to make equal angles.

144. The last four exercises lead to the following important principle : **If two lines drawn in the same plane diverge to the same extent from one side of a third line, they point in the same direction, or are symparallel.** Make clear why it is necessary to say "in the same plane" and "from one side," by giving illustrations of lines that

are not symparallel because they fail to answer all the conditions. Write the principle given above, using the notation explained in 121, scholium.

145. Make a line A B, and choose any point C outside of the line. Draw through C a line symparallel to A B. Suggestion : Draw C A as a help line, and use the principle of 144.

146. Make four lines symparallel to a line A B drawn at pleasure. Will these lines be symparallel to each other ?

147. Make a three-sided figure. Through each corner draw a line symparallel to the opposite side.

148. Make a quadrilateral with its opposite sides parallel. Such a figure is called a **parallelogram.**

149. Can you discover any truth about the angles of a parallelogram without actual measurement ? Record the principle in the " Summary."

150. Make a triangle A B C, and through the vertex C (Fig 33) draw D E ↑↑ A B. Select pairs of equal angles, telling in each case how you know that the angles are equal. What can you say about the sum of the angles with vertices at C ? What can you say about the sum of the angles of the triangle ? Compare the result obtained

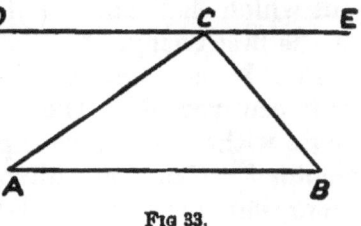
Fig 33.

here with the result of your experiments in 106. Notice the difference in the methods of reaching the conclusion. In this case the conclusion is reached after studying one triangle, in 106 after experimenting with several triangles ; in this case what we know about the angles depends upon the nature of angles, while in 106 our knowledge is based upon our measurements. We might have some doubt as to whether the truths learned about the few triangles with which we experimented in 106 would apply to all triangles ; but we are absolutely sure that the principle discovered in

150 applies to all triangles, because it is based on the nature of angles and lines. In this case the principle is **proved**; in 106 it is merely **tested** by experiment. Record the principle in the "Summary."

151. Make a triangle A B C with $\angle$ B A C = 70°, and $\angle$ A B C = 43°. How large is $\angle$ A C B? Why? Through each vertex, A, B, and C, in turn draw a line parallel to the opposite side. Four new triangles will thus be formed. Give the value of each angle in degrees without using your protractor.

152. One angle of a triangle is 35°, another 47°; find the third angle. If two angles of a triangle are 48° and 95° respectively, what is the third angle?

153. Can you make a triangle whose angles are 90°, 40°, and 60°? Why? Can you make a triangle with two obtuse angles in it? Can you make a triangle two of whose angles shall be right angles?

154. Make any triangle. Can you make a second triangle which shall have two angles like two of the first triangle, but which shall have its third angle unlike the third angle of the first triangle?

155. If the angles of Fig. 34 are two angles of a triangle, how can you find the third angle without using your protractor? Can you think of more than one way of solving this problem?

156. Two angles of a triangle are said to determine the third angle. Give illustrations of the meaning of this statement.

157. If one angle of a parallelogram is equal to $\angle$ x (Fig. 34), how can you find the other three angles without constructing the parallelogram? See 149.

158. One angle of a parallelogram is 72°. What are the other angles?

159. Find all the angles of a parallelogram, if one angle

is 90°; if one angle is 120°; if one angle is $\angle y$ (Fig. 34).

160. How many angles of a parallelogram are required to determine the remaining angles? Put your answer in the form of a principle, and record it in the "Summary."

161. In the parallelograms whose angles you have studied thus far, can you discover a principle that applies always to the adjacent angles?

162. Extend side C D of the parallelogram A B C D to R. What can you say about the respective directions of the sides of $\angle$ R D A and $\angle$ D A B? What follows from the directions of the sides of the angles?

163. Make an angle A B C. At some point P on A B *between* A and B draw (on the opposite side of A B from B C) a line, P R, that shall diverge from P B as much as B C diverges from B P. What is the geometrical name for angles R P B and P B C (121 Scholium)? Which of the following sentences is true? P R ↑↑ B C or P R ↑↓ B C? Write a translation of the correct sentence. Is P R parallel to B C?

164. Make a line A B and choose a point C outside of A B. Draw a line through C parallel to A B, without prolonging your help line A C beyond C. Will the line be antiparallel or symparallel to A B?

165. Repeat the problem of 164 three times, placing A, B, and C in different positions each time.

166. The last four exercises lead to the following principle: **If two lines are crossed by a third in such a way that the alternate interior or alternate exterior angles are equal, the two lines are parallel.**

167. Combining the principles of 144 and 166, we can say that two lines are parallel, if a third line crosses them in such a way as to make any pair of corresponding angles equal, or any pair of alternate interior or alternate exterior angles equal.

168. A practical way of drawing parallels by making the

corresponding angles equal is to slide a triangular piece of wood or stiff paper along a ruler until the edge passes through any desired point. For instance, in Fig. 35 it is desired to draw a line through P parallel to A B. Place your ruler in such a position that one edge, x, of your triangle will have the direction of A B; then slide the triangle along the ruler until the same edge, x, passes through P, and draw M N. The ruler represents a line crossing M N and A B in such a way as to make the corresponding angles equal.

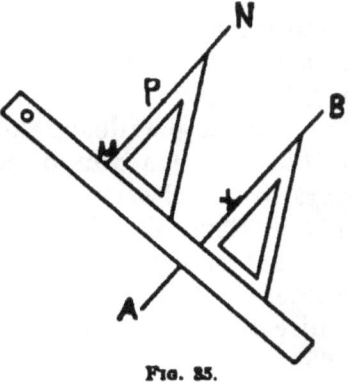

Fig. 35.

169. Make a parallelogram by the method described in 168.

170. Mark off six equal distances on a line A B; then, by the method of 168, draw parallels through the points of division.

171. Make a triangle; then through each vertex draw a line parallel to the opposite side, using the method of 168.

SECTION VIII. TRIANGLES.

172. Make a triangle A B C (Fig. 36). The sign for triangle is △, and for triangles △. What is the difference between △ A B C and ∠ A B C? Give as full an answer as you can to this question, naming all the respects in which they differ. Are they alike in any respect?

173. What is the difference between △ A B C and △ B A C? What is the difference between ∠ A B C and ∠ B A C? In how many ways can you name the △ A B C,

and in how many ways can you name the ∠ A B C, using the letters A, B, C each time?

174. Write a short sketch of the △ A B C, telling how many parts it has, what it is by nature, what you have

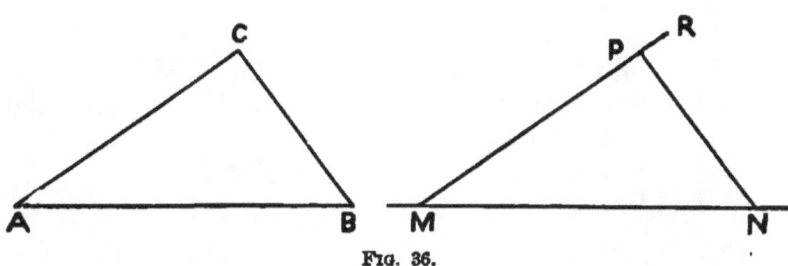

FIG. 36.

learned about its angles, and what fact about its sides you can infer from 124. Also write all you can about ∠ A B C. In writing these sketches, take great pains with your English.

175. After making △ A B C (Fig. 36), mark off, on a long line, M N equal in length to A B; then make ∠ N M R equal to ∠ B A C; then mark off on M R a length M P = A C; and, finally, join P N, thus completing △ M N P. Notice that you made the lines M N and M P and the angle N M R to match the corresponding parts of the △ A B C, but that, after taking these steps, you had no control over the line P N, because you were obliged to draw it between two fixed points P and N. With your protractor compare ∠ M P N with ∠ A C B, and ∠ M N P with ∠ A B C; and, with your compasses, compare P N with C B, recording your results by means of the sign = or the sign ≠ (unequal) as the case may be. If you find that P N is not equal to C B, write P N ≠ C B, but, if you find that P N is equal to C B, write P N = C B, etc.

Record your work from the beginning according to the model which follows, taking pains to record each step *as you take it*, not waiting to give your description after all the work is done. Put your figure at the *upper right* corner of your page, unless it is so large that it requires the whole width of the page.

    I. I make $\triangle$ A B C.   Position of figure.
    II. I make M N = A B.
    III. I make $\angle$ N M R = $\angle$ B A C.
    IV. I make M P = A C.
    V. I join P with N, thus completing $\triangle$ M N P.
    VI. I find that $\angle$ M P N $\neq$ or = $\angle$ A C B (using the proper sign).
    VII. I find that $\angle$ M N B $\neq$ or = $\angle$ A B C.
    VIII. I find that P N $\neq$ or = C B.

176. Repeat the experiment of 175 three times, using triangles of new shapes each time, and recording your work according to the above model.

177. Do your experiments in 175 and 176 seem to point to any truth about triangles? Take care to express the principle in the simplest and most accurate wording possible.

178. If your work was neatly and carefully done in 175 and 176, you probably were able to discover the principle involved. But, at the best, you have merely *tested* a principle in a few cases; you have not *proved* it by showing that it depends upon the nature of triangles, or upon principles already proved. Let us try to *prove* the principle that two triangles must be equal in all respects, if they have two sides and the included angle alike respectively.

There is a very great difference at the very start between testing and proving a principle. In testing a principle, everything depends upon the accuracy of the mechanical work; if there is the slightest error in making the angles or the lines, the test is a failure; but in proving a principle the drawing is simply to aid the imagination, by making it unnecessary to carry a great many lines and points in the mind at once; the accuracy of the drawing makes no real difference in the results. although an inaccurate drawing often misleads the mind; the mind should be constantly on the imaginary lines and angles represented, not on the lines and angles drawn.

Imagine, then, two triangles, A B C and M N P, to have two sides, A B and A C, and the included angle, B A C, of the first equal respectively to two sides, M N and M P, and the included angle M N P of the second. You are to try to prove that the remaining parts, B C and the angles A B C and A C B of the first triangle, must be equal to the remaining parts, N P, and the angles M N P and M P N of the second triangle. Imagine △ M N P placed on △ A B C, and try to prove that the triangles must exactly fit one another. How many points can you place before △ M N P becomes fixed in position? After placing M on A, are you free to place N wherever you please? In how many places could you put it? Which one will you choose? After placing M and N in two definite positions, in how many positions could you place P? Remember that P must be somewhere on the same plane surface with △ A B C, since you imagined the two triangles placed together. Have you any reason for thinking that P, if on the same side of A B with C, must be on the line A C? What is your reason? If you can prove that it must be somewhere on A C, do you know exactly where it must be? Remember that M is fixed at A. How do you know its exact position on A C? Why is there now no doubt as to the position of N P? See 61, Axiom. If you can find satisfactory answers to the questions asked above, and if you can explain your answers by reasons that depend upon the nature of lines and angles or upon principles already proved (and you ought to be able to do this with ease), you have proved the principle that **two triangles are necessarily equal in all respects, if they have two sides and the included angle alike respectively.**

179. By the aid of this principle of 178 we are able to decide without measurement that two lines must be equal under certain circumstances, and we have found a new way for determining the equality of two angles without protract-

or or compasses; for two lines or two angles must be equal, if they are corresponding parts of triangles that are known to be equal.

180. The principle of 178 is of great practical value in enabling us to find the distance from one point to another that cannot be seen from the first point because it is hidden by a building, or wall, or some intervening obstacle. Suppose you wish to find the distance from a stake on one side of a house to a stake on the opposite side of the house. Represent the position of the first stake by a point A, and that of the second stake by a point B. Then choose a point C from which you can see both A and B, and measure the angle A C B.

To measure the angle A C B, you need an enlarged protractor mounted on a tripod. If possible, get some surveyor to show you his theodolite, and to explain to you how he uses it. If you divide a semicircle having a radius of four or five inches into degrees, and mount it on a tripod such as you would use for a camera, you can measure angles with sufficient accuracy for your purposes. A stick about a foot long, provided with sights and pivoted at the centre of the protractor, will take the place of the surveyor's telescope. From the centre of the protractor suspend a plumb-line. Placing the tripod so that the point C is exactly below the centre of the protractor, sight the point A, and note the division at which the line of sight crosses the protractor; then turn the stick until you can sight point B, and again note the division of the protractor crossed by the line of sight; from the two readings of the protractor estimate $\angle$ A C B. By "sighting point A" is meant sighting a staff held in a vertical position at A. Next, measure the distances C A and C B. If, with the measurements that you have made, you construct a second triangle in a level place where there are no obstacles, you will have a triangle exactly like $\triangle$ A B C, so that you can measure the line corresponding to A B.

181. If at any time you should be without a mounted

protractor, can you think of any way in which you could use a piece of string and a stake to measure ∠ A C B in the last problem?

182. Describe the measurements that you would take to find the distance across a pond.

183. In practice it is often inconvenient to find a suitable place in which to make a triangle as large as the original triangle. Therefore, instead of making the lines C A and C B as long as they were found to be, we might make them one-half as long, or one-tenth as long, or we might use any convenient fraction of the actual length. In such cases the line found would be the same fraction of the original line A B. We might go still further, and use a centimeter or an inch to represent several feet, and we then could draw a plan of the ground on paper. The process just described is called **"drawing to scale."** In drawing to scale you cannot be too careful; a slight error makes a great difference; for instance, if one centimeter stands for ten feet, an error of a millimeter in your plan would make an error of one foot on the ground. In choosing your scale it is well to reduce the lines as little as the size of your paper will allow.

How long a line can you represent on your paper, if one centimeter represents ten feet? How long a line, if one inch represents ten feet?

184. Draw lines on your paper representing lengths of 150 ft., 161 ft., 84 ft., 26½ ft., using the scale 1 cm. = 10 ft.

185. Draw lines representing lengths of 75 ft., 63 ft., 110 ft., 87 ft., using the scale 1 in. = 20 ft.

186. Draw a triangle, and find the lengths of the sides of the triangle which it represents, if the scale is 1 cm. = 10 ft. ; if the scale is 1 in. = 10 ft.

187. If you are to represent lines between 300 ft. and 400 ft., what scale will you choose for your paper?

188. When the scale is 1 cm. = 5 ft., what do lines 7.5 cm., 24 cm., 63 mm., and 2 m. represent?

189. If the scale used is ¼ in. = 3 ft., how long are the lines which are represented by 2¼ in., 3⅛ in., 7¼ in. ?

190. Two sides, A C and B C, of a triangular field were found to be 150.75 ft. and 200 ft. respectively, while the included angle was 70°. Draw a plan of the field, using the scale 1 cm. = 15 ft., and calculate the length of A B.

191. Choose two points, A and B, in different, but adjoining, school-rooms in such a way that one point cannot be seen from the other, but that both can be seen by a person standing in the doorway between the rooms. Explain what measurements you would take to find the distance A B; actually take the measurements, choose a scale suited to the size of your paper, draw a plan, and find the length of the line A B.

NOTE.—A great many problems of this kind can be given with profit. It is important, in the first few cases at least, that the actual measurements be taken by the pupil. Later, if it is thought best, measurements can be taken from city plans that will enable the pupils to find the shortest distance between well-known points. In all cases neat and accurate work should be insisted upon. It is well to have the pupils select scales adapted to the paper used, and to have the same problem solved by different scales.

192. In 178 we found that two triangles would be alike necessarily, if they were known to have two sides and the *included* angle alike respectively. Let us see if we cannot leave out the word *included* and show that two triangles must be alike in all respects, if they have two sides and an angle alike

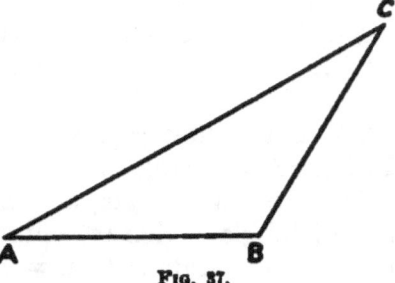

Fig. 37.

respectively. Make a triangle, A B C, taking heed to make A C longer than B C (Fig. 37). Make another triangle R S T having—

1st. R S = A C ;
2d. ∠ S R T = ∠ C A B ;
3d. S T = C B.

After taking the first two steps, are you obliged to take the third step, which will complete your triangle, in one way, or do you have a choice of ways? Make the following statement true by crossing out the words that need to be crossed out : **Two triangles are or are not necessarily equal, if they have two sides and the angle opposite the smaller side alike respectively.**

193. Making use of the same triangle (A B C, Fig. 37), make a triangle M N P having—

1st. N P = B C ;
2d. ∠ P N M = ∠ C B A ;
3d. P M = C A.

You remember that C A was purposely made longer than B C. After taking the first two steps, can you take the third step in more than one way? What words do you think ought to be crossed out in the following statement : **Two triangles are or are not necessarily equal in all respects, if they have two sides and the angle opposite the greater of the two sides alike respectively?** If, then, we leave out the word *included* in the principle of 178, what must we put in its place?

194. Make three different triangles. Make three new triangles by taking in each triangle of the first set two sides and an angle opposite the greater of the two sides chosen.

195. Make as many triangles as you can by taking in each triangle of the first three triangles drawn in 194 two sides and the angle opposite the smaller of the two sides chosen.

SECTION IX. REVIEW.

196. What are parallel lines?

197. If a line on the floor is parallel to a line on the ceiling, could you move one to the position of the other

without changing its direction? What would the moving line generate?

198. Can you draw two parallel lines so that a plane surface cannot be found which will contain both lines? Upon what axiom does your answer depend?

199. If a line on the floor is not parallel to a line on the ceiling, can you move one line to the position of the other without changing its direction? What would the moving line generate in reaching the position of the second line?

200. Draw several pairs of lines that would not meet within the limits of your paper, if extended. Estimate the angles between them, pair by pair, and record your estimate; then, by the method of 168, draw a parallel to one line of each pair that shall cross the second line, measure the angle, and record the true value and the error. In doing this exercise take care to have some of the lines make obtuse angles and some acute angles.

201. What is an oblique angle?

202. What is a geometrical proof?

203. One angle of a parallelogram is 50°; what are the other angles?

204. One angle of a parallelogram is twice another; what are the angles?

205. Extend side A B of the triangle A B C to R. Try to show that $\angle$ C B R is equal to the sum of two angles of the triangle, $\angle$ B A C and $\angle$ A C B.

206. Repeat the problem of 205, extending a line at each vertex of the triangle, and showing what angles of the triangle are needed to make each exterior angle formed.

207. One acute angle of a right triangle is 70°; what is the other?

208. Make a triangle A B C, and make another like it in as many ways as you can, by principles already established.

209. Can two triangles have two sides and an angle respectively alike and still be unequal?

SECTION X. ISOSCELES TRIANGLES.

210. Make a triangle A B C with A C = B C. Such a triangle is said to have equal legs, and for that reason is called an **isosceles** triangle, a name that in Greek means "with equal legs." The third side, in this case A B, is called the base of the triangle; the isosceles triangle is generally thought of as standing on this base, the legs meeting in a point which is called *the* vertex. In triangles that are not isosceles any side may be thought of as the base, and no one vertex is singled out as *the* vertex. In an isosceles triangle M P N, if P is the vertex, what is the base, and what are the legs?

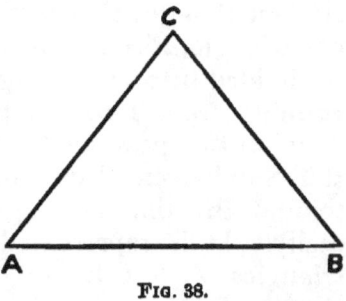

FIG. 38.

211. Triangle R S T has R T = S T (Fig. 39). To the eye ∠ R seems like ∠ S. Recall the tests for equal angles thus far discovered, and decide whether any one of them will prove the angles equal. Do the sides point the same

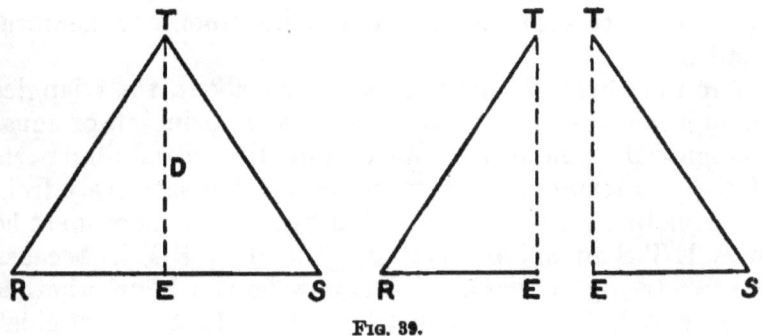

FIG. 39.

way respectively? Do the sides point in opposite directions respectively? Are they supplements of the same angle? Study the directions of the sides of the angles carefully, and be able to give a reason for your answer, whether yes or no. If all these tests fail, there is still one more chance for proving the angles equal, because in 179 it was found that two angles must be alike, if they are corresponding angles of triangles that can be proved equal. At first thought this test seems of little use, because only one triangle, R S T, is mentioned here; but the triangle can be divided into two triangles in a great many ways by lines running from T to different points in the base. Choose from all the possible lines the one that divides the angle R T S in halves. You will be obliged to use your protractor to find this line, although the imagination will supply it easily. It is represented by T D E in the figure. Two triangles, $\triangle$ R T E and $\triangle$ S T E, are now formed; to aid the imagination it may be well to draw them apart from each other (Fig. 39), or to make a paper isosceles triangle and to cut it along the line corresponding to T D E.

In comparing $\triangle$ R T E and E T S we know—

1st, that T R = T S from the nature of isosceles triangles;

2d, that T E = T E from the nature of straight lines;

3d, that $\angle$ R T E = $\angle$ E T S because each is one-half of R T S: so that our knowledge of these two triangles is based upon the nature of the figures and not upon our measurements.

Are the three things that we know about the triangles enough to prove them equal? Upon what principle of equal triangles do you rely? Write out the remaining parts of the two triangles that must be equal respectively from the principle stated in 179. You know that there must be in $\triangle$ E T S an angle equal to $\angle$ R of $\triangle$ E T R, because the two triangles are alike; but how do you know whether $\angle$ S or $\angle$ E is the one equal to $\angle$ R? In this particular case there can be no doubt, because $\angle$ E evidently is much larger than $\angle$ R; but it is not safe to rely upon the eye in such matters. A method of selecting the corresponding angles and lines that can never fail is based upon the fact that **the corresponding angles are opposite the sides that have been proved equal, and the corresponding sides are opposite the angles that have been proved equal.** In this case E T has been proved equal to E T : opposite E T in $\triangle$ E T R is $\angle$ R ; opposite E T in $\triangle$ E T S is $\angle$ S ; $\angle$ R and $\angle$ S, therefore, are corresponding angles of the two triangles, and are equal by 179. Select the other corresponding parts by the same kind of test. You cannot acquire too much skill in selecting corresponding parts of figures.

212. The knowledge that we have gained in 211 about isosceles triangles gives us a fifth test for detecting equal angles without protractor or compasses. We now know that **the base angles of an isosceles triangle must be equal.**

213. In showing that $\angle$ R = $\angle$ S (Fig. 39) we discovered a fact that is of even greater importance, namely, that $\angle$ R E T = $\angle$ T E S. What kind of angle must each one be ? Why ? State the fact as a principle relating to the bisector of the vertical angle of an isosceles triangle. A principle that is discovered while we are trying to prove another principle is called a **corollary.**

214. There is a second corollary resulting from the proof in 211 about the point E, in which the bisector of the vertical angle strikes the base. State it, and record the statement decided by the class to be the best.

215. Can you without your protractor tell the number of degrees in any angle of $\triangle$ E T R ?

216. If $\angle$ R of $\triangle$ R T S (Fig. 39) is 70°, what are the angles at S and T ? Why ?

217. What are the base angles of an isosceles triangle, if the vertical angle is 70° ? if it is 60° ? if it is 90° ?

218. If the base angle of an isosceles triangle is twice the vertical angle, what is the value of each angle of the triangle ?

219. Make a triangle with all three sides, or legs, equal.

What can you say about its angles? Why? Such a triangle is called an **equilateral** triangle. Why?

220. The vertical angle of an isosceles triangle is twice the base angle; can you tell the degrees in each angle?

221. In Fig. 40, △ A B C is an isosceles triangle; ∠ C B D formed by extending the base A B is 140°. What are the angles of △ A B C?

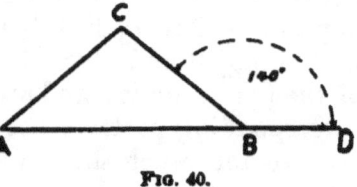

Fig. 40.

222. In Fig. 41 △ M N S is an isosceles triangle whose base angle, M, is 60°. M S is extended so that S P = MS. Can you name in degrees all the angles of the figure? What is ∠ P N M?

223. What would be the value of ∠ P N M (Fig. 41) if we should take ∠ M = 50°? What, if ∠ M = 70°? if ∠ M = x. Can you discover any principle about ∠ P N M?.

224. When two lines form a right angle they are said to be **perpendicular** to each other. The sign for perpendicular is ⊥.

What examples of perpendicular lines have we had in Section X.? Give examples of perpendicular lines in the room about you.

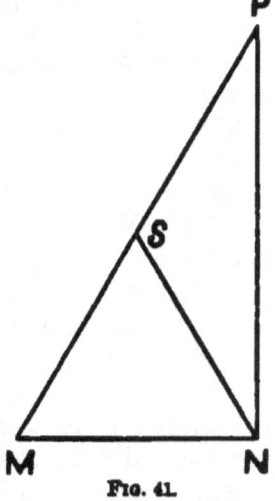

Fig. 41.

225. Can you use the principle of 223 to draw a perpendicular to a line A B at B? Suggestion: Choose a point C corresponding to S in Fig. 41, and outline the necessary isosceles triangles.

226. Draw six different lines, and draw perpendiculars to them, using the method of 225. Reduce the construction lines as much as you can in the last two or three cases. If you thoroughly understand the principle, you can so reduce the help lines that they will hardly be noticed.

227. In proving the fact about the base angles of an isosceles triangle (211), it was necessary for you to imagine the vertical angle bisected, or else to use the protractor to bisect it. For our proof the actual bisecting line was not needed (why?); but, as a matter of practical construction, it is important to be able to bisect an angle, and a construction problem is not looked upon as solved in Geometry, unless the ruler and compasses are the only tools used. Can you bisect an angle with ruler and compasses only? Can you prove that your method is a sound one?

If unable to bisect an angle for yourselves, study the figure given here (Fig. 42), and try to give reasons for the statements made.

1st. With your compasses make C A = C B, using C as centre.

2d. With A and B as centres, make A K = B K, putting K below A B.

3d. Join C K and A B, crossing each other at D.

1. ∠ C A B = ∠ C B A. Why?
2. ∠ D A K = ∠ D B K. Why?
3. ∠ C A K = ∠ C B K. Why?
4. △ C A K = △ C B K. Why?
5. ∠ D C A = ∠ D C B. Why?

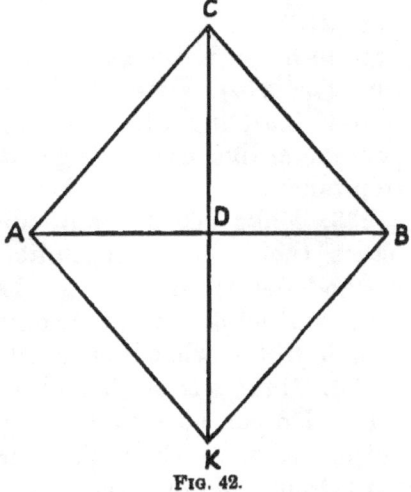

FIG. 42.

If the five statements made can be shown to be necessarily true, there can be no doubt that the line C D K bisects the angle A C B.

228. Make six different angles, and bisect them. In the first four cases go through the steps carefully, giving all the reasons which prove in the end that the angle is bisected. In the last two cases reduce the help lines to those which are absolutely necessary.

229. Draw a triangle and bisect each angle.

230. Draw a triangle, A B C. Extend A B, B C, and C A. Bisect the external angles formed.

231. Could K have been between C and the line A B in 227? Study Fig. 43, where C A = C B, and D A = D B, before answering. Are enough facts known about the figure to prove that △ C A D = △ C B D? If so, what follows about angles A C D and B C D?

232. Now that we know how to bisect an angle without a protractor, we know also how to draw a perpendicular to a line (see 213) in a new way, and how to bisect a line at the same time (see 214). Give an illustration of bisecting a line by a perpendicular line, explaining every step taken.

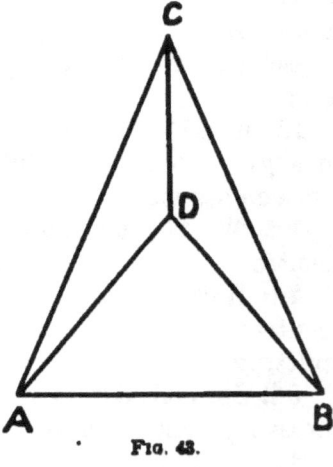

Fig. 43.

233. Make six different lines in various positions and bisect them with perpendicular lines according to the method described in 232. Leave to the imagination all lines and points not absolutely necessary for your work, but be able to tell what lines need to be imagined.

234. Make a triangle and find the middle point of each side. Do you notice anything peculiar about the bisecting perpendiculars when they are prolonged? Experiment with four different triangles, making at least one of the triangles an obtuse-angled triangle, and note whether or not there is the same peculiarity always.

235. Make an equilateral triangle, and draw semi-circumferences without the triangle on each side as a diameter.

236. Draw a circumference about each side of a square, A B C D, as a diameter.

237. Can you, by the method of 232, draw a perpendicular to the line A B at any point of A B except the midde point? Give three illustrations.

238. Draw a line, A B, choose a point P (not the centre) on A B, and draw a perpendicular to A B at P, first by the method used in 226, and then by the method just explained.

239. Can you use the method of 232 to draw a perpendicular to A B from a point P outside of A B? Give three illustrations.

240. Draw a line, A B, and choose a point, P, above and beyond the end of A B. How will you proceed to draw a perpendicular to A B from P in this case?

241. Draw a triangle, A B C. From each vertex draw a perpendicular to the opposite side. Do you notice any peculiarity of the three perpendiculars? Repeat the experiment three or four times, taking an obtuse-angled triangle twice.

242. The lines drawn in 241 from the vertices of a triangle perpendicular to the opposite sides are called the **altitudes** of the triangle, because they represent the heights of the triangle as it stands on its different bases. Does the altitude of a triangle ever join the vertex with the middle of the base? Is the altitude ever as long as a side of the triangle?

243. With the aid of the isosceles triangle, wholly or partly drawn, lines and angles can be bisected, perpendiculars to lines can be drawn, and equal angles may be discovered. You cannot, therefore, become too familiar with what has been shown about the isosceles triangle. Try to answer the following test questions without reference to your note-books. If you are obliged to look up the answers, you will know that you have not sufficiently mastered the subject, and that you have not been studying in the right way.

1st. If two isosceles triangles have the same base, what can you say about the line that joins their vertices?

2d. How many truths can you give about the line that bisects the vertical angle of an isosceles triangle?

3d. If you should draw three different isosceles triangles on the same base, could you say anything about the positions of their vertices?

4th. If a base angle of an isosceles triangle is 20°, what are the remaining angles? If the vertical angle is 20°, what are the remaining angles?

244. Make a straight angle. How does this differ from making a straight line? Bisect and also quarter the straight angle.

245. Make an angle of 60° without your protractor; then divide it into four equal angles. Divide a semi-circumference into twelve equal parts.

246. Divide a line, A B, into four equal parts; into eight equal parts. Can you divide it into three equal parts with the aid of isosceles triangles?

SECTION XI. TRIANGLES—Continued.

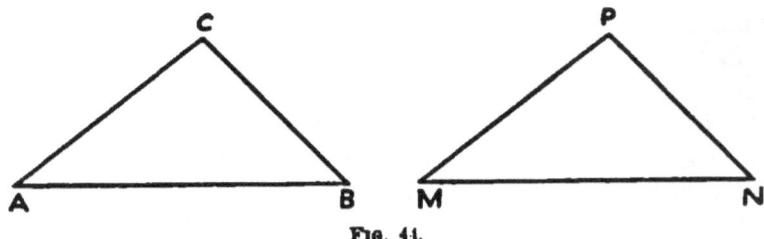

Fig. 44.

247. Make a triangle, A B C (Fig. 44). Make a second triangle, M N P, as follows:
  1st. Make M N = A B.
  2d. Make ∠ M = ∠ A.
  3d. Make ∠ N = ∠ B, thus completing △ M N P.

Over what parts of the triangle M N P have you had control, and over what parts have you had no control?

Can you think of any reason why $\angle$ P must be equal to $\angle$ C? Review 178, and then show that $\triangle$ M N P, if imagined to be placed upon $\triangle$ A B C, would exactly coincide with $\triangle$ A B C. Describe with great care the placing of $\triangle$ M N P, and show, by giving reasons, what must be the position of the rest of the triangle after you have placed two of its points.

248. Can you complete $\triangle$ M N P by making,
1st. M N = A B;
2d. $\angle$ M = $\angle$ A;
3d. $\angle$ P = $\angle$ C?

You will at first be puzzled to decide where to put the vertex of the angle like $\angle$ C; but a little ingenuity will enable you to find the right position without measuring A C, B C, or $\angle$ B. See 155, if you need a hint.

Will this triangle necessarily be equal to the triangle A B C? Show, by placing $\triangle$ M N P upon $\triangle$ A B C, whether or not it will exactly coincide with it.

249. As a result of 247 and 248 make the following statement true, and record the corrected statement in the summary:

Two triangles are or are not necessarily equal in all respects, if two angles and a side of one are equal respectively to two angles and a corresponding side of the other.

250. Can you construct a triangle that shall have a side A B, and its angles equal to those of $\triangle$ A B C, and yet be unequal to $\triangle$ A B C, because the side equal to A B is not in a corresponding position? How many such triangles can you construct?

251. Make any two angles whose sum will not amount to a straight angle; call these angles x and y; also make any line a. Construct as many triangles as you can that shall have the angles x and y and the line a. Can you make two unequal triangles that shall have the side a between the angles x and y?

252. A person wishing to find the distance from his house on the sea-shore to an island, sighted from his house, represented by point A in Fig. 45, a prominent object, B, on the island, and also a place, C, on land, which

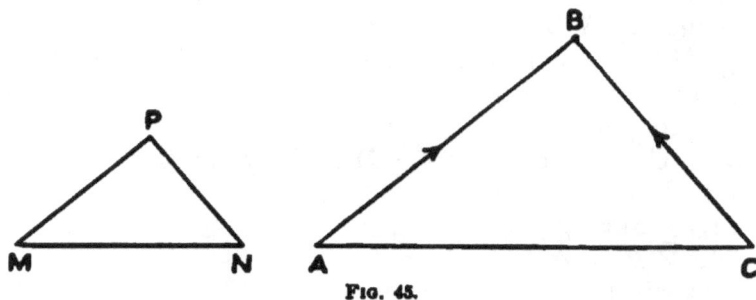

Fig. 45.

could be easily reached, and from which B could be seen; after noting the angle B A C, he measured A C, and sighted B and A from C, noting the angle A C B. With these measurements before him, he constructed on paper the triangle M N P, having $\angle M = \angle A$, and $\angle N = \angle C$. How did he estimate the distance A B?

253. Let the class calculate the distance from the school-house to a tree which can be seen from two windows of the room, making the proper measurements with as few suggestions as possible from the teacher. The measurements may be taken in the class, and the calculations may be made later, if the time seems too short.

254. What additional measurement would be needed to determine the height of the tree?

255. Describe and take the measurements that would be necessary to determine the length of a ladder that would reach to a particular branch on the same tree, if the ladder rests on the ground five feet from the tree.

256. A ship captain sights a light-house at an angle of 70° from his course. After sailing three miles on the same course, he sights it again at an angle of 100° from his course. How far was he from the light-house each

time? How far, when he passed nearest to the lighthouse?

257. Observe two points on a house or tree, one of which is directly over the other. Take the measurements that will enable you to calculate the distance between the two points without leaving the room.

258. Can you think of any reason why you can not use the method used in 256 to find the distance to a star?

259. A man, standing on the bank of a stream at a point A, sees at an elevation of 60° the top of a tree, which stands at a point B on the opposite shore; he walks away from the shore on the line B A for 80 ft.; he then sees the top of the same tree at an elevation of 30°. Draw a plan to scale, and find the height of the tree and the width of the river.

260. Construct a right triangle with one leg 3 in. and the adjacent acute angle 50°.

261. Construct a triangle having one angle 30°, another 65°, and the side joining the vertices of these angles $3\frac{1}{2}$ in.

262. Construct a triangle, A B C, with $\angle$ A = 70°, $\angle$ B = 60°, and side B C = 2 in.

263. How would you find the distance between two islands without leaving the main land?

264. In measuring a triangular field, A B C, a man found the side A B to be 300 yds.; he found the angle A B C to be 95°; when he reached C, he was not satisfied with his measurement of B C. Instead of retracing his steps to measure B C, he sighted A from C, finding the angle A C B to be 40°. How did this help him to find B C? What was the length of B C? of A C?

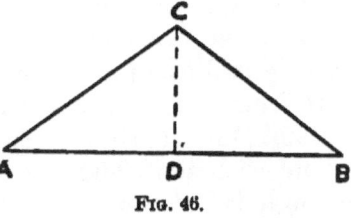

Fig. 46.

265. Construct a triangle, A B C, by making A B any desired length, and by making $\angle$ B = $\angle$ A. Will this triangle have equal legs, A C and B C? If you determine your answer by the appearance of the figure, or by testing the lengths with your com-

passes, you will doubtless answer that A C and B C are equal. But is there any reason in the nature of things why they must be equal? Let us investigate. Imagine $\angle$ A C B bisected by a line, C D, and, to aid the imagination, draw the bisector C D by the method of 227. Examine carefully △ A D C and B D C. Point out what parts of the first have their equals in the second. Are there enough equal parts to make the triangles necessarily equal in all respects? What principle of equal triangles do you rely upon? After satisfying yourselves that the triangles are equal, arrange your work as follows:

In △ A D C and △ B D C it is known,
1st, that $\angle$ A = $\angle$ B (by construction);
2d, that $\angle$ A C D = $\angle$ B C D (each $\angle$ = ½ of $\angle$ A C B);
3d, that C D = C D (same line).
∴ △ A B C = △ B C D.
(Quote principle that you rely upon here.)
∴ A C = B C (give reason here).
∴ △ A C B is isosceles.

∴ **A triangle must be isosceles, if its base angles are equal.**

266. Why could we not have said at once in 265 that AD = B D, and that $\angle$ C D A = $\angle$ C D B, thus enabling us to use a different principle for testing the equality of the triangles?

267. Make a triangle with its base angles 60°. What must the third angle be? What kind of triangle is it?

268. From a ship that was sailing in a straight course, A B, a light was sighted at an angle of 40° with the course A B; after the ship had gone four miles, the same light was sighted at an angle of 70° with B A. How far was the ship from the light at the first observation?

269. Make a right triangle, A B C, with right angle at C and with acute angle A equal to 60°. How many degrees in angle B? Draw a line, C K, in such a way that $\angle$ K C B = $\angle$ K B C. Study the figure thus formed, and select all the angles whose value in degrees you know without the aid

of a protractor; also select any lines that you can show to be alike without measurement. Do you discover any truth relating to the longest and shortest lines of a right triangle one of whose angles is 60°? Note the distance of K from A, B, and C. Compare what you note in this case with what you learned in 225. K is on A B.

270. Repeat the experiment made in 269, this time making $\angle A = 70°$. Make $\angle K C B = \angle K B C$ as before. Give the value of all the angles formed. Can you say the same thing about K that you did in 269?

271. Make a right triangle, A B C, without knowing the exact value of acute angle A. Draw C K as before so that $\angle K C B = \angle K B C$. Calling $\angle A = \angle x$, can you name all the angles formed by means of $\angle x$? Can you show that what you learned about K in 269 and 270 still holds true?

272. If with the middle point of the longest side or **hypotenuse** of a right triangle as a centre, you describe a circumference, will it pass through the vertices of the right triangle? Why? The principle involved in this problem plays an important part in later problems, so that it is well to record it, and to make special effort to remember it. It may be stated thus: **The middle point of the hypotenuse of a right triangle is the same distance from the three vertices of the right triangle.**

273. If, when walking on a straight level road, A B, you sight from A a tree across a field at an angle of 70° with A B, at what angle from B A will you have to sight the tree to know that you have walked a distance equal to that from A to the tree?

274. Wishing to cut down a tall tree, a man found that it was of great importance to know how far the tree would reach when felled. Having no instruments with him, he fastened two straight sticks together at an angle of 45°. How could he get an angle of 45° degrees with only a piece of string and a stake to work with? Can you describe how he used the simple instrument to find the height of

the tree and the distance wanted? Why did he choose the angle 45°?

275. How many ways of recognizing equal angles without measurement have you learned thus far? Give them all.

276. How many ways of recognizing equal lines without measurement do you know?

277. Give the three tests for recognizing equal triangles.

278. Make a triangle (A B C, Fig. 47). Make M N =

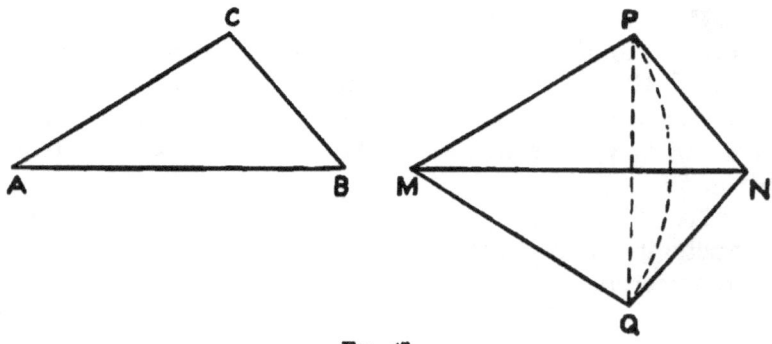

FIG. 47.

A B. With M as centre and with A C as radius describe an arc. With N as centre and B C as radius describe an arc cutting the first arc at P and Q. Complete the triangles M P N and M Q N. Clearly these triangles have the same three sides that triangle A B C has. Is there any reason for thinking that △ M N P has the same angles as △ M N Q? If only one angle is the same in both, the triangles must be equal in all respects. Why? Let us aim to show that ∠ M P N = ∠ M Q N. Join P Q. What kind of triangle is △ P M Q? Why?

∴ ∠ M P Q = ∠ ...... Why?
Likewise ∠ N P Q = ∠ ...... Why?
∴ ∠ M P N = ∠ ...... Why?
∴ △ M P N = △ M Q N. Why?

Should you place △ A B C in the plane of the paper with

A B upon its equal M N, where would C have to fall? Why? Hence △ A B C = △ M N P = △ M Q N. Hence a new principle for recognizing equal triangles: **Two triangles are equal in all respects, if the three sides of one are equal to the three sides of the other respectively.**

279. In Fig. 47 can you tell the number of degrees in the angle between M N and P Q? Give reasons.

280. If, in △ A B C, A B is 7 in. long, and A C is 5 in. long, how short can B C be? How long can it be? What reasons can you give for your answers?

SECTION XII. QUADRILATERALS.

281. Construct quadrilaterals with your compasses and ruler as follows:

1st. A quadrilateral no two of whose sides are parallel. Such a quadrilateral is called a **trapezium**.

2d. A quadrilateral only two of whose sides are parallel. This figure is a **trapezoid**.

3d. A quadrilateral whose opposite sides are parallel. What is this figure called?

282. If one line crosses two others, what angles must be alike to make the lines parallel? Is there more than one pair of angles whose equality will make the lines parallel?

283. How many degrees must there be in the sum of the four angles of a quadrilateral? See 106–110.

284. Can you make a trapezium with two of its sides alike? with three of its sides alike? with all four of its sides alike? In the last case, draw a line joining two opposite corners, dividing the quadrilateral into two triangles, and, from your study of these triangles, decide whether the figure may be a trapezium or must be another kind of quadrilateral.

285. Can you make a trapezium with two angles alike?

with three angles alike? with four angles alike? Can you make a trapezium with three of its angles 60° each? Why? Can there be three angles of 70° in a trapezium?

286. Can you make a trapezoid with two sides alike? with three sides alike? with two angles alike? with three angles alike?

287. Draw a trapezoid, A B C D, making A B and C D antiparallel. It is possible to fix the value of $\angle$ D from the value of $\angle$ A. Do you see how? If $\angle$ A = 70°, what is $\angle$ D? If $\angle$ B = 25°, what is $\angle$ C? Does $\angle$ A help you to find $\angle$ C?

288. If, in making a trapezoid A B C D, you first make $\angle$ A = 70° and $\angle$ B = 40°, can you make $\angle$s C and D of any size that you please, or are they fixed in value? Give a reason for your answer.

289. How many angles of a parallelogram can you make before all of the angles are fixed in value?

290. The most interesting of the quadrilaterals are the parallelograms, of which there are several varieties. Those with their sides all alike and with their angles 90° each are called **squares**.

Those that have their sides alike but their angles oblique are called **rhombuses**. Those that have right angles but adjacent sides unequal are called **rectangles**. Those that have oblique angles and adjacent sides unequal are called **rhomboids**.

291. Construct models of each kind of parallelogram, describing the steps *in the order* in which you take them and *at the time* at which you take them.

292. Select examples of the different kinds of parallelogram from the school-room. What kind of parallelogram is a sheet of writing-paper? a diamond? a window-pane?

293. Can you make a rhomboid with three of its angles alike? Why?

294. Is any proof needed to show that the opposite sides of a parallelogram are parallel, pair by pair? Why?

295. Can you draw a parallelogram whose opposite sides

are not equal? Does your answer to this need proof? Why?

296. Can you see any reason why a parallelogram should not be defined in the first place as a quadrilateral with its opposite sides *parallel and equal?*

NOTE.—It is important at this point to make clear, if possible, the danger of imposing too many conditions upon lines and angles, and also, in the same line of thought, to impress upon the pupil the importance of drawing the figures exactly according to the conditions, except in rare cases. If the condition imposed is that the lines be made parallel, they should be made parallel, and not equal, although it may be possible to prove that they are parallel when made equal. The subject should not be left until, by varied illustration, some impression is produced upon the pupil. It will generally be found necessary to return to the subject at short intervals. Abundant opportunity for testing the pupil's understanding of the subject may be found in questions about the line bisecting the vertical angle of an isosceles triangle.

297. If a parallelogram really cannot be drawn with its opposite sides unequal, there must be some reason for the fact based upon the nature of lines and angles, or upon principles already established. Draw an angle, D A B; complete a parallelogram by making D C ↑↑ A B and B C ↑↑ A D, taking pains to make A B and A D unequal. Draw a line, D B, joining two opposite vertices (the line is called a **diagonal**), and study the triangles thus formed. It should be a simple matter for you to prove △ A B D = △ B C D. Arrange your work in a vertical line, putting no more than one fact and its reason on a line. Follow this model:

I. I make D C ↑↑ A B.

II. I make B C ↑↑ A D, completing a parallelogram. To prove A B = D C and A D = B C.

III.. Draw diagonal D B and compare △A B D and D B C.

IV. ∠ .... = ∠ .... (Insert reason.)

V. $\angle$ .... = $\angle$ .... (Reason.)
VI. Side .... = Side .... (Reason.)
VII. ∴ $\triangle$ .... = $\triangle$ .... (Reason.)
VIII. ∴, etc.

298. It may help you to understand 297 to cut a parallelogram something like A B C D from stiff paper, and to cut the parallelogram into two parts along the diagonal D B. Can you place the two triangles formed by cutting the paper parallelogram so that they will not form a parallelogram? Is a quadrilateral whose diagonal divides it into two equal triangles necessarily a parallelogram?

299. Can you place the two paper triangles of 298 in more than one position so that they will form a parallelogram?

300. Does it make any difference which diagonal you draw? Will the parallelogram be divided into two equal triangles in each case?

301. How does a rhomboid differ from a rhombus?

302. How does a rhombus differ from a square?

303. How does a rectangle differ from a square?

304. Notice that, although squares, rhombuses, and rectangles are not rhomboids, they differ from the rhomboid merely in having special characteristics in *addition* to those of the rhomboid. Hence any general principle that is proved about the rhomboid is applicable to the rhombus, the square, and the rectangle; but it is dangerous to assume that a principle proved true of a square, a rhombus, or a rectangle can be applied to a rhomboid, because the principle may depend upon the special characteristics of the former figures. Therefore, when trying to prove a principle that will apply to all parallelograms, be careful to draw a rhomboid for illustration, or to have a rhomboid in your imagination.

305. Into what kind of triangles does a diagonal divide a square? a rectangle? a rhombus?

306. Can you construct a parallelogram with one angle 90° and another 80°? Why?

307. If the adjacent sides of a parallelogram are equal, is there any doubt as to what kind of parallelogram the figure is?

308. Can you make a rhomboid with one side 3 in., another 2 in., and a diagonal 5½ in.? Why?

309. Make a rhomboid, A B C D (Fig. 48). What precautions must you take to be sure that the figure is a rhomboid? Draw both diagonals, A C and B D, crossing at M.

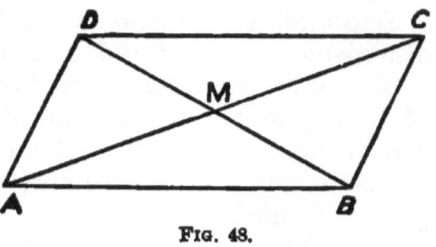

FIG. 48.

1st. Make a record in a column as in 297 of all the pairs of equal angles that you can find in the figure, giving a reason in each case.

2d. Select pairs of lines that you can declare to be equal, because of principles already established, giving the principle in each case.

3d. Select four pairs of equal triangles in the figure, stating the proper principle for each case.

4th. From the equal triangles show that certain lines in the figure must be equal to each other that were not known before to be equal. What can you say of the point M? How do the diagonals divide each other? Write and record a principle stating the fact that you have just learned about the diagonals of a parallelogram.

310. Draw a square, a rhombus, and a rectangle, and notice how their diagonals divide each other. Do the diagonals ever cross at right angles? Can you account for their crossing at right angles by the principles that enabled you to draw a perpendicular to a line?

311. Do the diagonals ever cross at a point which is the same distance from all four corners? You have already learned a principle that should enable you to answer this question; can you recall it?

312. What kinds of parallelogram can you **inscribe** in a

circumference? A figure is **inscribed** in a circumference when the circumference passes through all the vertices of the figure.

312. On smooth, stiff card-board draw two squares with sides 1¾ in. long. In the first square draw one diagonal. In the second square draw the lines represented in the second figure of the plate on the next page. Cut the squares into seven pieces, numbered as in the diagram.

Select all the pairs of equal lines in the second square; also give the number of degrees in every angle formed.

313. With the seven pieces formed by cutting the squares of 312 a great variety of interesting figures can be formed. For instance, by placing them as in the fourth figure of the plate, an old-fashioned chair like the third figure of the plate is formed.

314. Try to form figures like those given in the next three plates. Each time you must use all seven pieces. I have seen over three hundred figures formed with these seven pieces, many of them bearing quaint resemblances to common objects. With squares of wood, celluloid, or ivory, divided in the manner outlined, you may get much amusement by attempting new forms.

315. Arrange the seven pieces in such a way as to form a rectangle; a rhomboid; a square. Notice the side of the square formed. Which of the original lines forms its side? How much larger is the new square than the old one? Is a method of finding a square twice as large as a given square suggested?

316. If you put pieces 4, 5, and 7 together so that they form a square, how will that square be related to piece 3? What part of the original square will it be?

317. Write, for review, short sketches of the square, rhombus, and rectangle, following as a model the following sketch of the rhomboid: The rhomboid is a quadrilateral which has two pairs of parallel sides, its adjacent sides unequal, and no one of its angles equal to 90°. Each diagonal divides the rhomboid into two equal triangles; the two

CONSTRUCTIONAL GEOMETRY 61

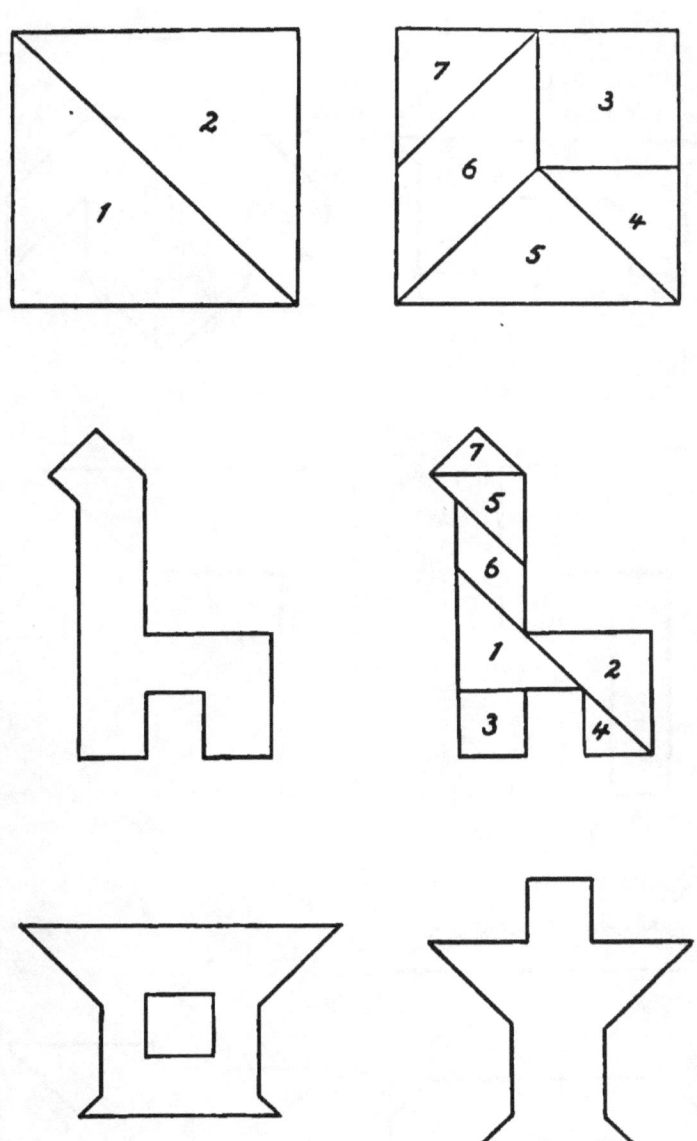

62 ELEMENTARY AND

CONSTRUCTIONAL GEOMETRY 63

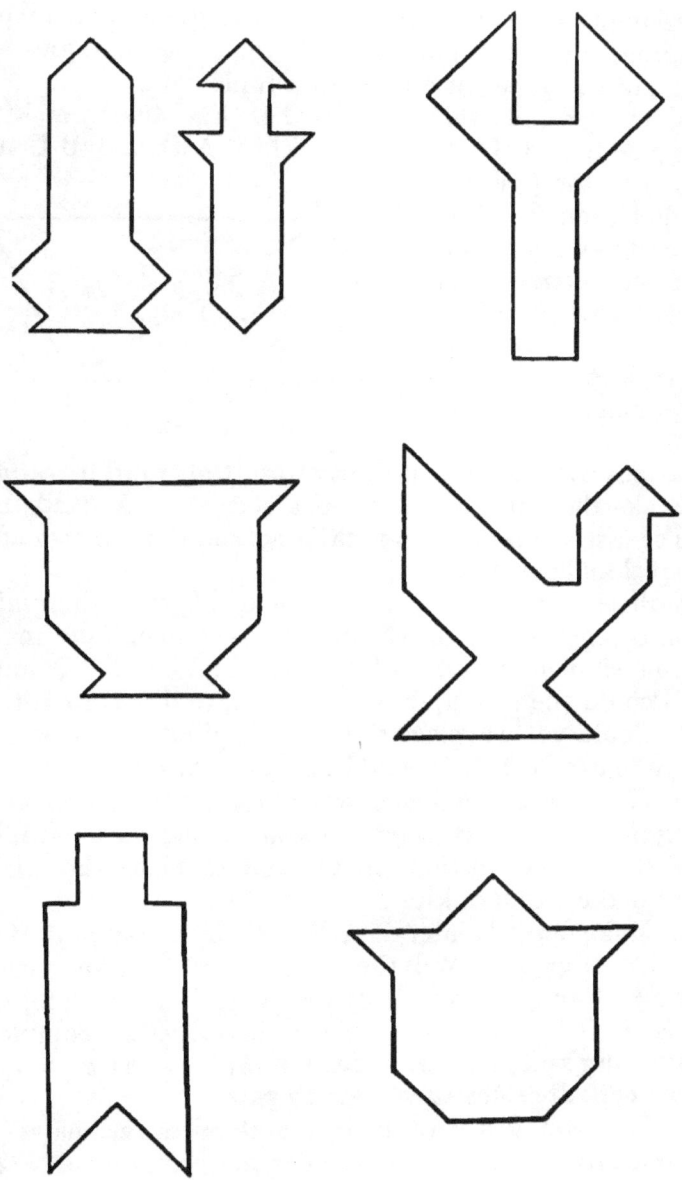

diagonals bisect each other, and divide the rhomboid into four triangles equal pair by pair. One angle of a rhomboid determines all the angles, because the opposite angles are equal, and the adjacent angles are supplementary.

318. Draw a quadrilateral, A B C D (Fig. 49), by making D C ↑↑ A B and D C = A B. When A D and B C are joined, will the figure be a parallelogram? Draw B D, and see if you cannot select enough equal parts to prove △ A B D = △ B D C. Arrange your work neatly, as already explained to you in 297. Draw as many conclusions as you can; and, as a result of your investigation, make the following statement correct : **A quadrilateral is or is not necessarily a parallelogram, if it has two sides both equal and parallel.**

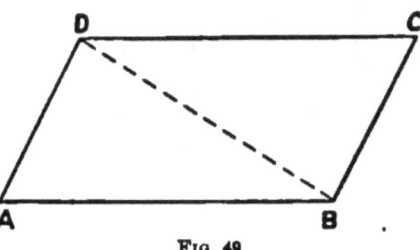

Fig. 49.

319. Place two *unequal* lines, M P and N Q, with their middle points together; make the lines cross at an oblique angle. Can you show what kind of quadrilateral M N P Q must be? Record the principle, when you have discovered it.

320. Could you so place the lines M P and N Q of 319 that the figure M N P Q would be a rhombus?

321. If you wish to place two sticks so as to make a rectangular kite, what conditions must the sticks fulfil? Would the same conditions enable you to place the sticks so as to make a square kite?

322. Make a quadrilateral, A B C D, by making A B = D C and A D = B C. Will the quadrilateral have any peculiar form? Can you prove it by drawing, as a help line, the diagonal B D? As a result of your investigation complete the following sentence : **A quadrilateral ...... be a ...... if it has its opposite sides equal pair by pair**.

323. Can you draw a quadrilateral with two pairs of equal sides that does not come under the principle of 322?

SECTION XIII. DIVISION OF LINES.

324. Draw a **scalene** triangle, A B C (Fig. 50). A **scalene** triangle is one that has its sides unequal. A scalene triangle bears the same relation to triangles that a rhomboid does to parallelograms. It is the kind of triangle that should be drawn for illustration when no particular kind is specified.

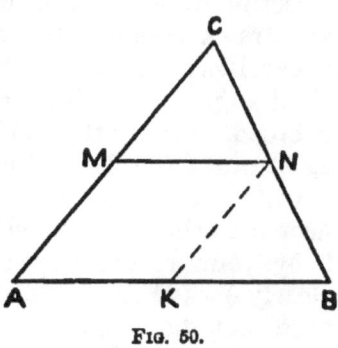

Fig. 50.

Find M, the middle of A C, and draw M N symparallel with A B. (Let your construction lines be light but visible.)

Draw a help line N K ↑↑ M A. (Note that help lines are always dotted lines.)

Study the figure thus formed, and write answers to the following questions, giving reasons in each case:

How many and what angles are equal to ∠ A?

What angle is equal to ∠ C? to ∠ B?

What kind of quadrilateral is A K N M?

What lines can be inferred to be equal from the nature of the quadrilateral A K N M?

How many angles of △ M N C can be found mated in △ K B N?

Is any side of △ M N C known to be equal to a side of △ K B N?

Is △ M N C equal to △ K B N?

If you have been able to answer these questions, and to support your answers with good reasons, you have proved that C N = N B and M N = K B; so that N is the middle point of B C, and M N is one-half of the base line A B.

325. The principle involved in 324 is one of the most important in Geometry. Stated in general language it is as follows:

A line drawn from the middle point of one side of a triangle parallel to the base bisects the other side, and is itself one-half of the base.

326. This principle suggests a new and simple method of bisecting lines by using common ruled paper, where the lines are drawn equal distances apart. A sheet of theme-paper is excellent for this purpose, because there is a red line at right angles to the blue ruled lines, and because the paper is broad. Along the red line number successive lines, 0, 1, 2, 3, etc. (Fig. 51). Place the line to be bisected on line 2 with one end at point 2; connect the other end, A, with point 0 by your ruler's edge; 1 B is clearly ½ of 2 A. The line 0 A need not be drawn; the same piece of paper, therefore, may be used for bisecting an indefinite number of lines.

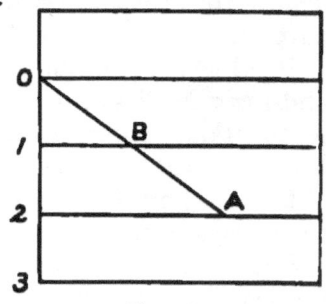

Fig. 51.

327. On what other lines between 0 and 12 could you place the line to be bisected, and on what lines respectively would the half-line appear? Bisect a line by placing it on three different lines, and compare your results. Can you see any principle to guide your choice of a line on which to place the line to be bisected?

328. Make a triangle, and then make another with its sides one-half as long, using the new method for getting the half-lines.

329. Make a quadrilateral—not a parallelogram—and find the middle points of its sides by the method of 326. Join the middle points of the sides in order by straight lines. What kind of quadrilateral will be formed? Can you account for its shape?

Suggestion: Draw one diagonal of the original quadrilateral, and apply what you learned in 325 and 318. The principle of 325 does not apply immediately, but it leads very easily to the right principle.

330. Make a rhomboid, and repeat the experiment of 329. Can you account for the shape of the new parallelogram formed? Can it be anything but a rhomboid?

331. Repeat the experiment of 329, using a rectangle. What kind of parallelogram will be formed by joining the middle points of the sides?

332. Try the same experiment, using a rhombus at first. Account for the shape of the second parallelogram formed.

333. Draw a trapezoid, A B C D (Fig. 52). Find M, the middle point of A D, draw M N ↑↑ A B, and study the figure formed, using as a help line the diagonal D B.

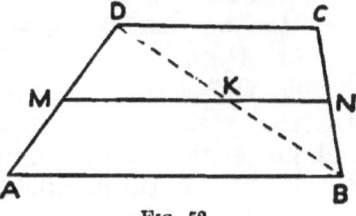

Fig. 52.

What reason have you for thinking K the middle of D B?

Compare M K with A B.

Have you any reason for thinking N to be the middle point of B C? Suggestion: Study the △ D B C. Compare K N with D C. Compare M N with A B and D C combined. The line M N is a famous line in the trapezoid. It is called **the median of** a trapezoid because it passes though the middle points of both legs. Notice carefully how the median was drawn. It was not drawn from middle point to middle point, but from one middle point parallel to the two bases, and was proved to pass through the middle point of the second leg, a characteristic from which it gets its name. Perhaps a more important characteristic is that it is half of the sum of the bases, or the average of the two bases. It is very important that you should become familiar with both characteristics and with the following principle: **The line drawn from the middle point of one leg of a trapezoid parallel to the bases bisects the other leg, and is itself one-half the sum of the bases.**

334. Make two lines, a and b, and find their average by the principle of 333, and with the aid of the ruled sheet of paper used in 326; also add the two lines a and b, and find one-half of the resulting line by 326. Compare the results.

335. What is the average width of a board which is 12 in. at one end and 9 in. at the other? Do this both by Geometry and by Arithmetic, and compare the results.

336. Find the average diameter of an ordinary wooden pail. Can you find the average diameter of a barrel? of a base-ball bat?

337. In a trapezoid, A B C D, find the median, if A B = $11\frac{1}{2}$ in. and C D = $5\frac{1}{4}$ in.

338. Find the base A B of a trapezoid, A B C D, if C D = 8 in., and the median, M N, = 12 in.; if C D = 11 in. and the median, M N, = $13\frac{1}{2}$ in.; if C D = $7\frac{1}{4}$ cm., and M N = $4\frac{1}{2}$ cm.

339. On a line, A D (Fig. 53), mark off five equal lengths with your compasses. Draw A B making an oblique angle with A D, and through the points of division on A D draw parallels to A B terminating in a line B C. Measure the upper two parallels and calculate the rest.

340. Could you calculate all the lines from measurements of the first and third lines, D C and E L? of the first and fourth lines, D C and G M?

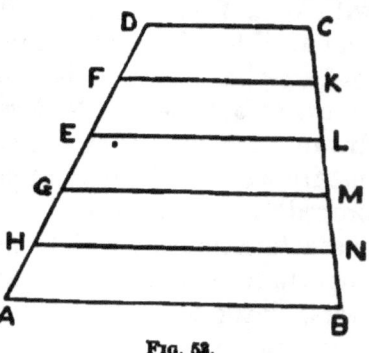

Fig. 53.

341. On a sheet of ruled paper draw two non-parallel lines in such a way that they shall intercept $\frac{1}{2}$ in. on the top line and 1 in. on the line below the top. Calculate the lengths intercepted on all the lines.

342. On a piece of theme-paper mark the points at which the red line crosses the blue lines, 0, 1, 2, 3, 4, 5, etc., in succession; draw from the point 0 a line, 0 Y, crossing obliquely the blue lines at points A, B, C, D, E, etc., respectively. If you rule the paper yourselves, take pains to have the lines drawn at equal distances and strictly parallel. Study the triangle

O 2 B. What can you say about the line 1 A ? the line O A ? If you give the name x to the line 1 A, what name will you give to the line 2 B ?

In the next place study the trapezoid 1 A C 3. The line 2 B is manifestly the median of the trapezoid, and, therefore, the average of the two bases. It follows that the line 3 C must be equal to $3x$, for $2x$ is the average between x and $3x$. By studying the different trapezoids in order, you will see that 4 D is four times as long as 1 A, or $4x$; that 5 E is five times as long as 1 A, or $5x$, etc. What would be the distance cut off on line number 10 by O Y ? on line number 16 ? Wherever you can, confirm your results by the middle-point principle of 326 and 327. Thus by studying △ O 4 D you see that 4 D ought to be twice as long as 2 B, or $4x$, the result already obtained.

343. Does it make any difference in the relative value of the lines cut off whether O Y diverges more or less from the red vertical line ?

344. If you should place a line one inch long on line number 10—placing one end of the inch line on the red line and joining the other end with O—what part of an inch would be found on line No. 1 ? on line No. 3 ? Have you ever seen a ruler for getting tenths of an inch arranged on this principle ? Cannot some one of the class obtain such a ruler for illustration in the class-room ?

345. How would you use these same ten lines to obtain $\frac{1}{100}$ of an inch ? $\frac{3}{100}$ of an inch ?

346. Compare the line 3 C, on line No. 3, with the line 5 E on line No. 5 ; also compare line 4 D with line 7 G. Then explain how you would get $\frac{2}{3}$ of a line, $\frac{4}{5}$ of a line, $\frac{5}{6}$ of a line, $\frac{7}{12}$ of a line by means of a sheet of ruled theme-paper.

347. Place any line, P Q, on the tenth line, with P at point 10. What part of P Q will be found between your ruler and point 7, if you place your ruler's edge from O to Q ?

348. Can you get $\frac{5}{6}$ of a line by using the same sheet of paper ? Explain your method. Draw a line, P Q, and get $\frac{3}{5}$ of it ; $\frac{7}{8}$ of it ; $2\frac{1}{2}$ of it.

349. Make a trefoil (Fig. 11), with the centres about one inch apart. Then make a figure just like it in shape —or similar to it—with all the lines ⅜ as long. Then enlarge the original figure by making the lines ⁴⁄₁ as long.

350. Make a rose window (Fig. 54) by putting the centres of the arches at points on a ¾ in. circumference which are ¾ of an inch apart. Reduce the figure by making the lines ¼ as long; also enlarge the figure by making the lines ⁴⁄₁ as long.

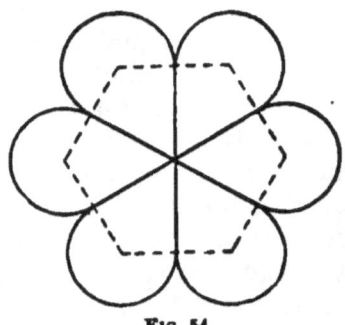

Fig. 54.

351. With only ten ruled lines can you find $\frac{1}{50}$ of a line? $\frac{7}{50}$ of a line? $\frac{1}{100}$ of a line?

352. Can you find ⁴⁄₁ of a line too long to put on your paper?

353. Find a line $\frac{1}{50}$ as long as your desk.

354. Can you find $\frac{3}{7}$ of a line with a sheet of paper containing only ten parallel lines?

355. It will be convenient for the future to refer to your ruled sheet of theme-paper as your **dividing tool**; for with it you can get almost any fraction of a line that you wish.

356. If, instead of putting point 0 on the top line, we had put it on line 5, say, and about half-way between the edges; and if we had numbered the lines above this zero line, 1, 2, 3, 4, 5, and the numbers below the zero line, 1, 2, 3, 4 5, 6, 7, etc.; and if we had drawn two oblique lines meeting at 0 across all the lines, we should have had a picture of an instrument called proportional dividers, a very useful instrument in reducing or enlarging a figure. The principle is the same as that of your dividing tool, but the tool is more convenient than yours in actual practice. Any line between the oblique lines on line 5 above the zero line is the same in length as the corresponding line on line 5 below the zero line, and this latter line we know to be ⁵⁄₇ as long as the length intercepted on line 7 by the oblique lines. Make a diagram to illustrate what has been said,

and see if you can follow it. Proportional dividers are dividers pivoted, not at one end as your dividers are, but somewhere between the ends at a point, corresponding to 0, that can be changed at pleasure by a sliding screw. If it is wished to reduce the lines of a figure to $\frac{3}{4}$ of their actual size, the dividers are pivoted so that the legs on one side of the pivot are $\frac{3}{4}$ as long as those on the other side. Now, when the longer legs are opened any distance, the shorter legs are opened only $\frac{3}{4}$ as far; so that all the reduced lines can be found with great rapidity without changing the pivot screw. The figures, pp. 61–63, were reduced from actual blocks by proportional dividers. If possible, secure a pair of proportional dividers to illustrate the important principle involved.

REVIEW.

357. Review from 281, and come into the class prepared to write a short sketch of any of the quadrilaterals thus far studied.

358. Do the diagonals of a trapezoid or a trapezium ever bisect each other?

359. What can be said of the diagonals of a rhombus that cannot be said of the diagonals of a rectangle, and *vice versa?*

360. A quadrilateral when cut along one diagonal, was found to be divided into two equal pieces; can you say what the shape of the quadrilateral was? Why?

361. Two sides of a quadrilateral were found to be equal and to be symparallel; what possible shapes could the figure have had?

362. In studying a certain quadrilateral a pupil noticed that the point where the diagonals crossed was the same distance from all the corners; do you know what quadrilateral he was studying?

363. If you have two lines of unequal length, how will your dividing tool enable you to tell whether the shorter is more or less than $\frac{7}{10}$ of the longer?

364. What principle of the trapezoid leads to the construction of your dividing tool?

365. With a pair of proportional dividers how would you reduce the figures on the plates following 314 so that the lines would be ⅔ as long?

366. If on side A B of △ A B C a point, P, is chosen ¼ of the distance from C to A, and if from P a line, P Q, is drawn symparallel to the base A B, what part of A B will P Q be? What part of C B will C Q be?

SECTION XIV. MULTIPLICATION OF LINES. AREAS.

367. We have already learned how to add lines and to subtract lines; in certain cases, with the aid of the dividing tool, we can divide one line by another; for the line intercepted on the ninth parallel by O Y of the dividing tool will contain the line intercepted on the fourth parallel by O Y 2¼ times. The question naturally arises whether or not two lines can be multiplied together.

Draw a line, M N (Fig. 55), and with your dividing tool find M Q = ⅓ M N. Place the line M Q at right angles to M N, and complete the rectangle M N S Q. Now divide M N into nine equal parts with your dividing tool. How many of these parts will M Q contain? Why? Through the points of division in M N draw parallels to M Q, and through the points of division in M Q draw parallels to M N. The small figures formed will be squares. Why? How many of them?

How many squares would there have been, had you made M Q = ⅔ of M N? How many squares in the figure M T R Q? If each one of the divisions of M N is called "b," each one of the little squares is called a "square b" or "b square" (written b^2). It is easy to see from experiments with different rectangles in Fig. 55 that any rectangle contains as many "square b's" as there are units in the product obtained by multiplying the number of b's in the base by the number of b's in the side. Thus the rectangle T N S R, with two b's in the base and five b's in the side, contains two lines five or ten "square b's," written $10b^2$.

In describing a floor we say that it contains 300 square feet—meaning by that that it covers as much surface as

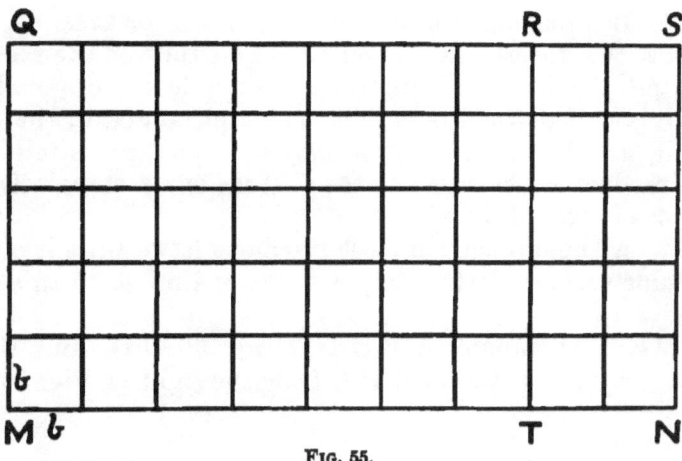

Fig. 55.

300 squares each 1 ft. long and 1 ft. wide would cover; and we find out how many square feet it covers by measuring the length and width in feet, and by multiplying the two results together. In the same way we measure M N and M Q in b's, and multiply the number of b's in M N by the number of b's in M Q, thus obtaining the number of "square b's" covered by the rectangle M N S Q.

368. It is, therefore, a simple step to say that the product of two lines represents a surface, bounded by a rectangle of which the two lines are adjacent sides. To multiply two lines, then, according to this understanding of the terms, we simply complete a rectangle with the two lines as adjacent sides. It is not necessary to form each time a pict-

ure of the little squares that correspond to b^2 in Fig. 55. On the contrary, it is better to think of the rectangular surface as a whole in thinking what the product of two lines means.

369. Draw three pairs of lines, and multiply the lines of each pair together. It will be well to review the method of drawing a perpendicular at the end of a line given in 225.

370. Draw three rectangles, and tell what lines must be multiplied together to produce them.

371. In problems relating to the surface, or **area,** of rectangles, the upright side is called the altitude of the rectangle, and the line on which the rectangle is supposed to stand is called the base. The principle stated in the preceding articles of this section may be expressed as follows: **The product of the base and the altitude of a rectangle is the surface of the rectangle.**

372. Although the simplest product of two lines is a rectangular surface, fortunately we are not limited to this surface.

Draw a rhomboid A B C D (Fig. 56 (a)), and draw B H $\perp$ A B. Cut the $\triangle$ B H C from the right of Fig. 56 (a)

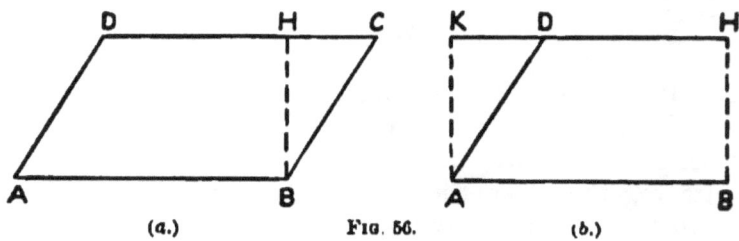

Fig. 56.

and place it to the left of A D, as in Fig. 56 (b). How do you know that B C will fit on A D, and that D K will be in a straight line with H D? How does the surface of the rectangle A B H K compare with the surface of the rhomboid A B C D? What product represents the rectangle? What product represents the rhomboid therefore? Illustrate this article by cutting a rhomboid out of paper.

373. The surface, or area, of the rhomboid is the same as the surface of the rectangle; but the rhomboid is not *equal* to the rectangle because one figure would not fit the other exactly. When two areas are the same the figures are said to be **equivalent**; a new sign, ⇌, is very commonly used to express this condition, although it is not a sign universally agreed upon. The sentence ▱ A B C D ⇌ ▭ A B H K is translated thus: The rhomboid A B C D is equivalent to the rectangle A B H K; and the meaning is that the figures have the same area.

374. What is the altitude of the rhomboid A B C D, Fig. 56 (a)? Translate this sentence: ▱ A B C D ⇌ A B × B H.

375. Make four rhomboids of different shapes, and draw the lines whose product gives the area in each case, recording the result according to the geometrical sentence in 374. Is there more than one answer in each case?

376. Multiply two lines so that the product shall be a rhomboid. Can you get more than one rhomboid for an answer? How many?

377. Can you multiply two lines and obtain the resulting rhomboid at once, without first drawing the rectangle of the two lines?

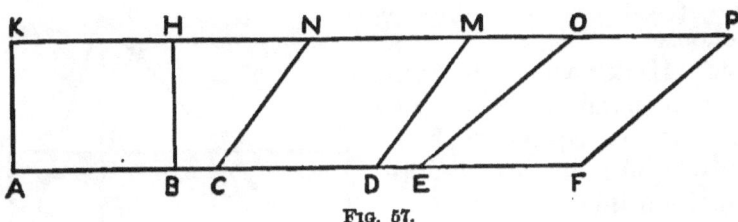

Fig. 57.

378. Make two parallel lines, A F and K P (Fig. 57). Make A B = C D = E F, and complete the rectangle A B H K and the two rhomboids C D M N and E F P O. Which of the three parallelograms has the largest surface? Why? Make clear what product represents each surface.

379. The sum of the bounding lines of a figure is called its **perimeter**. Do the three parallelograms of Fig. 57 have equal perimeters?

380. If a man has an acre of ground, will it make any difference to him, in the cost of fencing it, whether the field is in the shape of a rectangle or of a **rhomboid**? Which is the more economical shape?

381. Can you arrange a square foot of paper in the form of a rhomboid with a perimeter of fifty feet? Can you make the perimeter as long as you please without changing the surface of the paper? as short as you please?

382. In building a cistern of a fixed height to hold a certain amount of water, does it make any difference in the number of bricks used whether you make the bottom of the cistern in the shape of a rhomboid or of a rectangle?

383. Which lets in the more light, a diamond-shaped window pane or a rectangular window pane, if the perimeters of the two panes are the same?

Make two lines, a and b. Make a rhomboid equivalent to $a \times b$ which shall have a perimeter three times the sum of a and b.

384. If a farmer wants to have as little fencing as possible to build, and if he is obliged to have his field in the shape of some parallelogram, what parallelogram ought he to choose?

385. Under what conditions are two parallelograms with the same base equivalent? Illustrate your answer, and record the principle involved.

386. Can the product of two lines ever be a triangular surface? This is merely another way of asking whether a triangular surface can be changed into an equivalent rectangular surface. Draw a scalene triangle, A B C (Fig. 58).

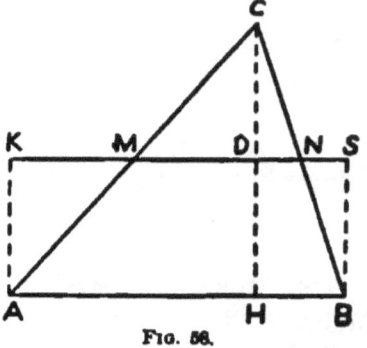

Fig. 58.

Find the middle point, M, of A C, and draw M N parallel to A B. Draw from C a line, C H ⊥ A B. (Show your construction lines.) What is C H called? What part of C H is C D? Why?

Erect perpendiculars to A B at A and B. Extend D M and D N until they meet these lines at K and S respectively. Show that △ C D N = △ S N B and that △ C M D = △ M K A.

Is △ A B C ⇌ ☐ A B S K? Why?

By what line must you multiply A B to produce △ A B C?

Follow the steps of this article with a paper triangle which you cut along the lines indicated.

387. Draw five different triangles, and draw side by side the lines whose product in each case represents the area of the triangle. In two of the cases make a base angle obtuse.

388. Make three pairs of unequal lines. Form three triangles whose surfaces will be represented by the products of the lines respectively taken pair by pair.

389. Is there only one triangle to be found equivalent to the product of two lines, *a* and *b*? If more than one, how many can be found? Do their perimeters differ?

390. Make two parallel lines; on the lower line, A D,

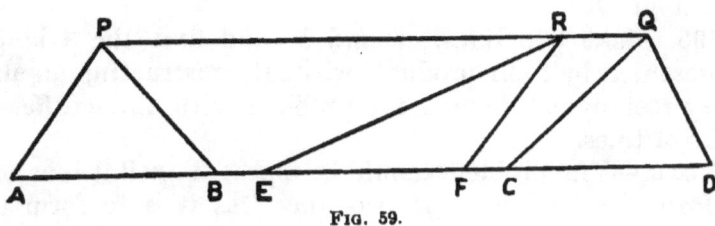

FIG. 59.

take three equal lengths, A B, E F, and C D; on the upper line take three points, P, R, and Q, at random; complete the triangles A B P, E F R, and C D Q. Which has the largest area? Which has the longest perimeter?

If you wish to find the triangle which has the shortest

perimeter possible, provided that the base equals A B and that the vertex is on P Q, what shape will you give the triangle?

391. Pin the ends of a short piece of elastic to a line A B at the points A and B; then form a triangle A B P with the elastic without stretching the elastic; draw through P a line, P Q, parallel to A B, and pull the vertex P of the elastic triangle along P Q. In this way you can mould the triangle A B P into a triangle having almost any desired shape. Does the area change in the process? When is the perimeter shortest? If the elastic could be stretched indefinitely, would there be any limit to the perimeter?

392. Can you mould the elastic triangle of 391 into a right triangle? an isosceles triangle? an equilateral triangle?

393. Learn by heart the following principle:

If the vertex of a triangle is moved parallel to the base, the area of the triangle remains unchanged as long as the base remains unchanged.

394. Make a triangle, A B C, and mould it into an equivalent triangle with one side twice as long as A C without changing the base A B.

395. Make a triangle, A B C, and mould it into an equivalent isosceles triangle; mould it also into a triangle with base angle 70°.

396. Make two lines, *a* and *b*, and find the triangle represented by their product, without constructing an auxiliary rectangle. Repeat the problem with three different pairs of lines.

NOTE.—The problem should be repeated until it is as easy to form the "triangle of two lines" as it is to form the "rectangle of two lines."

397. Make a pyramid with a square base. Draw the lines, five pairs, the sum of whose products, pair by pair, represents the total surface of the pyramid.

398. If three triangular tin boxes of equal height, with bottoms and tops shaped like the three triangles of Fig. 59,

were to be made, which box would hold the most sand? Would one box require more tin than another? Why? Can you draw pictures of the boxes? What is the geometrical name for the sides of the boxes? When boxes are made with triangular bottoms, what shape is usually chosen? Do you understand why?

SECTION XV. EQUIVALENT FIGURES. MOULDING OF AREAS.

399. The principle of 393 enables you to mould triangles and all plane figures of geometry almost as easily as if the figures were made of elastic or of clay. For instance, suppose that two triangles, △ A B C and △ B D E of Fig. 60, are of the same height, and are standing, side by side, on the same base line as in Fig. 60. They can be moulded

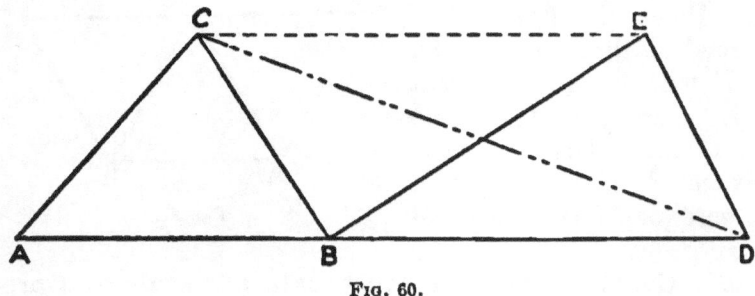

FIG. 60.

into one triangle, A C D, with perfect ease. Do you see why △ B C D is equivalent to △ B E D?

NOTE.—In this figure and in future figures, when figures are moulded, the lines along which vertices are moved will be indicated by light dotted lines, as all help lines should be indicated; the resulting lines of the remodelled figure will be indicated by double dotted lines. In your drawings follow the same practice, unless you are able to use different colors for the different kinds of lines.

400. Add together two triangles which have the same altitude, but different bases. Repeat with three pairs of triangles.

401. Can you add three triangles that have the same altitude ? Can you do this problem with only one double dotted line ?

402. Find an isosceles triangle equivalent to three triangles that have the same altitude.

403. Mould three triangles having the same altitude into a right triangle.

404. Make six equilateral triangles from one point, O, forming the six-sided figure, or **hexagon**, represented in Fig. 61. With a sharp knife cut the figure along the lines O A, O B, etc., *beginning at O* and not cutting quite up to points A, B, C, etc., except at point F. Place the figure thus formed so that A, B, C, D, E, F will form a straight line. Can you form one triangle as large as the hexagon ? How long will its base be ? How tall will it be ?

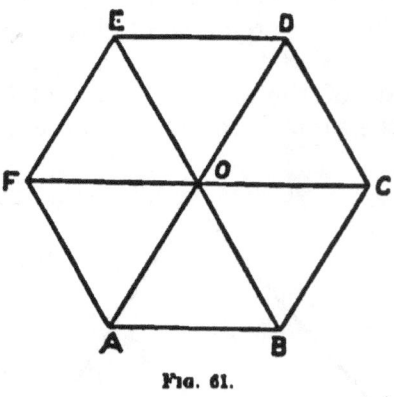

Fig. 61.

405. Could you mould a square into a triangle by a process similar to that of 404 ? a parallelogram ?

406. Make a quadrilateral, A B C D, Fig. 62. The problem is to mould this quadrilateral into an equivalent triangle. The principle of 393 suggests a method. Divide A B C D into two triangles by the diagonal B D. C D B can now be looked at as a triangle standing on the base D B with vertex at C. C can be moved parallel to D B as far as you wish to move it, and D C and B C will follow the

moving C just as the legs of the elastic triangle in 391 followed the moving vertex. When C reaches R, in the line of A B, △ C D B has been moulded into △ R D B, and the quadrilateral A B C D has been moulded into △ A R D.

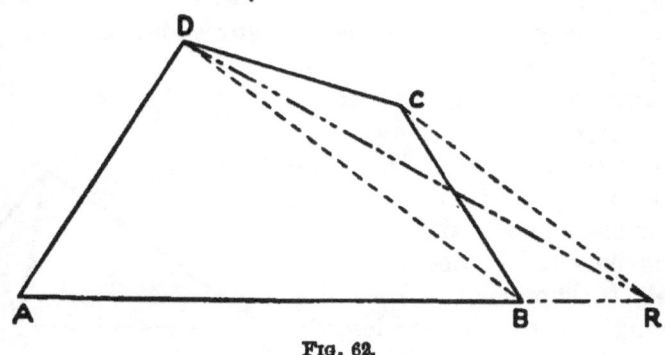

Fig. 62.

This process can be illustrated very prettily on a moulding-tray with clay or sand in the shape of a quadrilateral, and should be so illustrated if possible.

407. If you had moved C in a line antiparallel to D B, where would you have stopped the movement? Illustrate with a figure, using the same quadrilateral as in 406.

408. Draw the diagonal C A, Fig. 62, and mould the quadrilateral into a triangle in two ways, first moving D symparallel with C A, and in a second figure moving D antiparallel to C A.

409. Make a quadrilateral, and mould it into an isosceles triangle.

410. Make a quadrilateral, and mould it into a right triangle.

411. Can you reverse the process of 406, and mould a triangle A D R (Fig. 62) into a quadrilateral A B C D? Suggestion: Choose any point between A and R for B, and draw D B as a guide line, to which you must make R C parallel. If all the lines and points of Fig. 62 were blotted out except △ A R D, what would be your chances of reproducing quadrilateral A B C D?

412. Make a △ A B C and mould it into a trapezoid. What difficulty can you see in moulding a triangle into a parallelogram by this process?

413. In moulding a triangle, A B C, into a quadrilateral (Fig. 63), if, after choosing the point P between A and B for one vertex of your quadrilateral, you should move B parallel to P C, not stopping until you should reach A C extended, the figure formed would be a new triangle, A P Q. The base of △ A B C has been shortened and the altitude lengthened. Hence it is possible to shorten the base of a triangle to any desired length. Illustrate this article with a clay or sand triangle, pushing in the base to any desired point and noticing the increase in altitude.

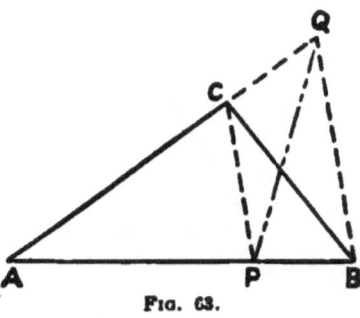

Fig. 63.

414. In 408 you changed one quadrilateral, A B C D, Fig. 62, into four triangles by using different guide lines, D B and C A, and by moving C and D first in one direction and then in the opposite direction. Place these four triangles on one line, side by side, and shorten all the bases to one length, three-fourths of the line A B. If your work is accurately done, where should the vertices of the raised triangles be? Why? Take pains to have this construction an almost perfect specimen of workmanship.

415. In 413 you shortened the base line just as much or as little as you pleased, but the vertex rose to a point that was out of your control. You, of course, knew that the more you shortened the base the higher the vertex rose, but you could not fix the level to which the vertex would rise, at least without further construction or arithmetical calculation. Can you raise a triangle to any desired altitude? Try to solve the problem by yourselves, but, if you

find it too hard, study Fig. 64, where △ A B C is the original triangle and H P is the desired altitude. M B is the guide line, parallel to which you draw C Q, to determine the length A Q of the shortened base. This problem is just the reverse of 413, and you can easily see that your construction is right by shortening the base line to A Q, and by noticing that you would move B along a line symparallel with Q C to

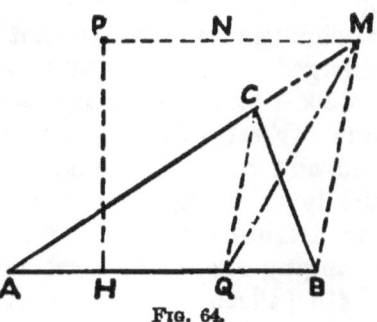

Fig. 64.

M. If Q C is the guide to B M in one construction, it is not hard to see that M B becomes the guide to C Q in the reverse construction.

416. Make three different triangles and raise the vertex to a level fixed upon beforehand.

417. In 415 we shortened the base by cutting off something at the right end, moving B in as far as Q. Try to solve the same problem by cutting off a length at the left end, thus pushing A in toward B. Suggestion: Extend B C until it crosses P M at N, Fig. 64, and draw a new guide line, N A. Compare the resulting base line with A Q of Fig. 64.

418. Make three triangles, and raise the vertices to any desired level by pushing the left end of the base toward the right end.

Practise changing triangles to triangles of greater altitude until it is a matter of indifference to you on which side of

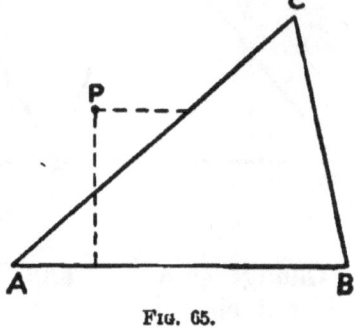

Fig. 65.

the triangle you draw your construction lines.

419. Can you now take the necessary steps to lower the

84   ELEMENTARY AND

vertex of a triangle without changing the area? Make a triangle, A B C, and lower the vertex to the level of P, Fig. 65. Do this twice: first, so that the position of A shall not be changed; second, so that the position of B shall not be changed. Compare the two bases obtained.

420. Place two triangles of different altitudes side by side. First, raise the lower one to the level of the other, and add the two triangles; second, lower the taller one to the level of the shorter, and add the two. Compare the two triangles resulting from addition in the two cases by changing one to the level of the other.

421. Place two triangles, one of which is a great deal taller than the other, side by side. Change them to a common level by raising one and lowering the other; and then add the results. In changing levels work from the left side of the left-hand triangle, and from the right side of the right-hand triangle, that the vertex where the triangles join may not be changed in the process.

422. Place three triangles of different heights side by side as in Fig. 66.

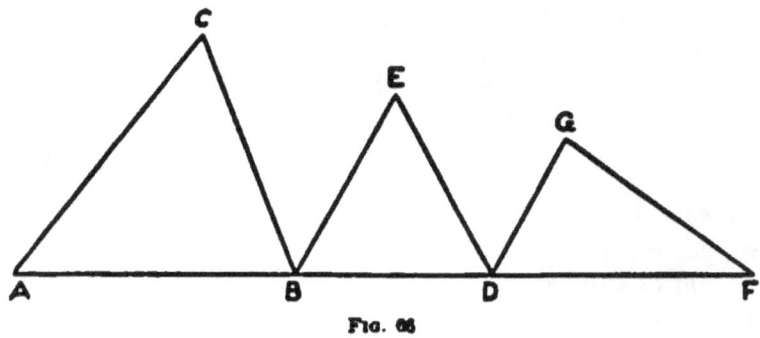

Fig. 66

Change △ A B C and D F G to the level of △ B D E without changing the positions of B and D. Add the resulting triangles. Aim to complete the problem in as few steps as possible, indicating your steps by the kind of line drawn.

423. Make a triangle and a quadrilateral that seem to you to have the same area. Change the quadrilateral to an equivalent triangle, and change this triangle to the level of the first triangle, and draw a triangle representing your error in estimation, if there be any error.

424. Make a quadrilateral free hand that seems equivalent to the sum of a triangle and a quadrilateral that you have drawn. Test your accuracy as in 423.

425. Change a pentagon into an equivalent triangle. Suggestion: First mould the pentagon into a quadrilateral, and then mould the quadrilateral into a triangle. The drawing is much less confused, if you work on different sides of the figure in taking the two steps.

426. Find a triangle equivalent to the sum of a pentagon and a quadrilateral.

427. Can you change a triangle into a pentagon?

428. Figures having six, seven, eight, or any number of sides can be changed to triangles by reducing the number of sides one at a time, just as the pentagon in 425 was changed to a quadrilateral, and the resulting quadrilateral was changed to a triangle. After two steps the lines become confused, unless a new figure is drawn.

All plane figures that are bounded by straight lines are called **polygons**. Special names have already been given to polygons of three, four, five, and six sides. What are they?

429. Can you give a rule for changing any polygon into an equivalent triangle, that will hold good no matter how many sides the polygon has?

430. Draw a circumference and divide it into twenty-four equal parts. Can you do this without a protractor? Form a polygon of twenty-four sides by joining the points of division. The polygon is inscribed in the circumference. What is meant by this statement? This polygon can be moulded into a triangle by a process much simpler than the one which you have described in 429; for, by joining the vertices with the centre of the circumference, you form twenty-four equal triangles. See 404, where a hexagon was

divided in this way. Since the twenty-four triangles have the same altitude, they can be changed into one triangle very easily. Describe the method. How long a base will the resulting triangle have? How tall will it be?

431. Describe how you would find a polygon of forty-eight sides inscribed in a circumference from the inscribed polygon of twenty-four sides of 430. How could you change this polygon into a triangle? How long would the base be?

432. In the same way that you got an inscribed polygon of 48 sides from one of 24 sides, you could get one of 96 sides from one of 48 sides, and then you could get polygons of 192, 384, 768 sides in succession, continuing the process until the polygons seemed to the eye to coincide with the circle. Each time the polygon could be moulded into a triangle having a base equal to the perimeter of the polygon and an altitude equal to the distance from the centre of the circumference to one of the sides of the polygon.

433. It is not difficult to see from 432 that the circle itself could be moulded into a triangle having a base equal to the circumference and an altitude equal to the radius. The difficulty is to find a straight line equal to the circumference. Make the following experiments and bring your results into the class: Roll a cylinder of wood about two inches in diameter in a straight line, making it turn ten times. By marking the lowest point of the cylinder at the start and the point on the table just below this point, it is a simple matter to find the distance rolled and the number of turns made. Compare the distance rolled with the diameter. Push a 28-inch bicycle until the wheel has made ten revolutions. Measure the distance covered and compare it with the diameter. Perform each experiment several times and record the results in a neat table.

How many times as long as the diameter do you find the circumference to be on the average?

434. Using the results of 433, change five different circles into essentially equivalent triangles.

435. Can you tell why it is a much more difficult problem to change a triangle into a circle? Suggestion: Think how much longer than the altitude the base of the triangle ought to be, to enable you to call the altitude of the triangle the radius of the equivalent circle.

436. Draw a particular triangle that you can turn into a circle with as much accuracy as the results of 433 allow.

SECTION XVI. PECULIARITIES OF SQUARES.

437. Make an isosceles right triangle. The problem is to change it into an equivalent right triangle with a leg equal to any chosen length, say P Q, longer than A B. One way of solving this problem is to raise the vertex by the method of 415, until the height A R is equal to P Q. Take the necessary steps and find the equivalent triangle A S R.

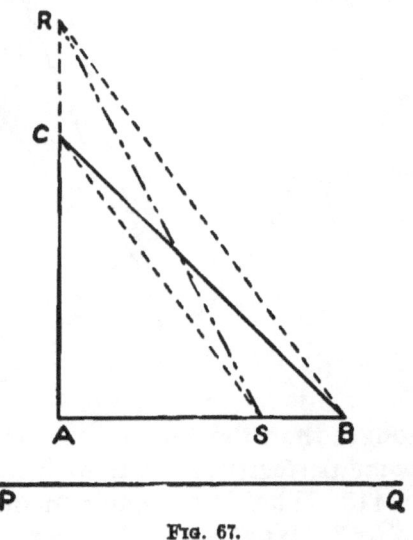

Fig. 67.

438. A much more roundabout way of solving the problem of 437—but for the present purposes a better way—is to proceed as follows: move C, Fig. 68, along a line parallel to B A, until it reaches a point E that is at a distance P Q from A, thus moulding △ A B C into △ A B E.

Now imagine △ A B E to be pivoted at A, and to be turned through a quarter of a revolution into the position of △ A C R. △ A C R ≎ △ A B C. Why? Now move C

parallel to A R to S, thus moulding △ A C R into △ A S R.
△ A C R ⌒ △ A S R. Why? Hence △ A S R ⌒ A B C;
since △ A S R has the desired leg A R = A E = P Q, the
problem is solved. Compare the result with the result of
the last article.

439. Using the method of 438, turn isosceles right triangles into right triangles with one leg of any chosen length

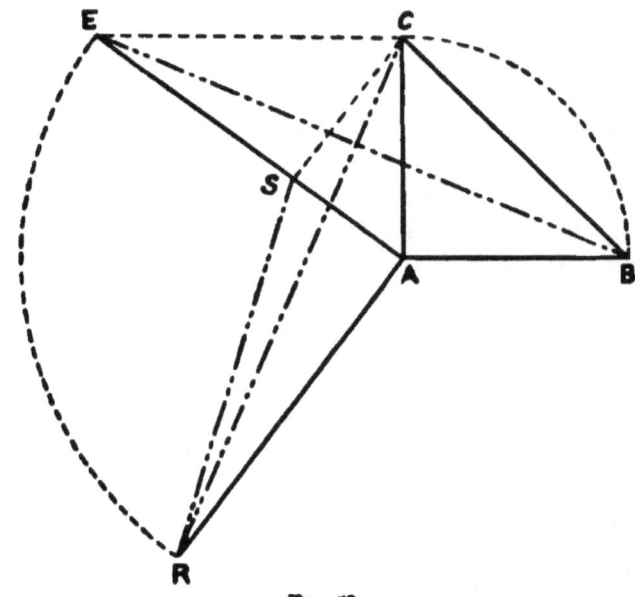

Fig. 68.

longer than the leg of the isosceles triangle, until you become perfectly familiar with the process.

440. What is the angle formed by S C and S A, Fig. 68?
Why? What kind of triangle is △ A C E?

441. Can you turn an isosceles right triangle A B C into
a right triangle A S R, Fig. 68, without actually drawing
lines B E and C R? Try three cases of this kind. The
lines should be imagined, though not drawn.

442. The isosceles right triangle A B C, Fig. 68, is one
half of a square, and the right triangle A S R is one half of

a rectangle. Draw a figure showing the square and rectangle meant. What is true of their areas? Why?

The process described in 438 gives you the means of moulding a square into a rectangle having one side any desired length longer than the side of the square. Do you see how?

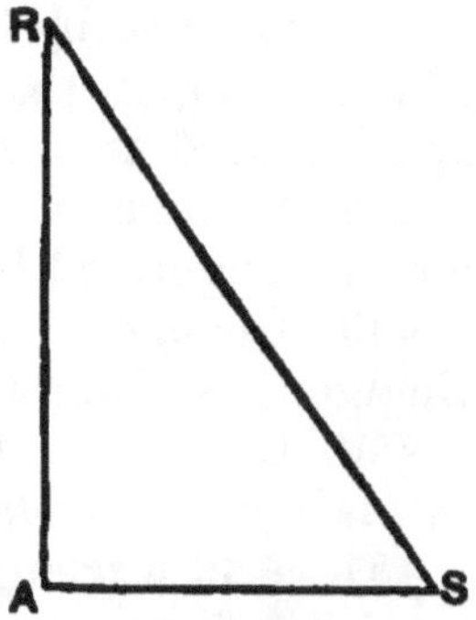

443. Turn four squares into rectangles.

444. Turn one square into four different rectangles.

445. In 437 and 438 we have seen how to turn an isosceles right triangle into a right triangle with unequal legs, in two different ways. Can a right triangle with unequal legs be changed to an isosceles right triangle? Try first to reverse the method of 437, and to change a right triangle A S R, Fig. 69, to an isosceles right triangle A B C by lowering the vertex R to the proper level C, so that the extended base A B shall be equal to the reduced height A C. How will you take the first step? Why is it difficult to draw a guide line, parallel to which R may be moved?

446. In the next place, try to reverse the method of 438. To help the imagination, place triangle A S R (Fig. 70) in a position something like that of triangle A S R in Fig. 68.

The problem still is to find C, to which to move the vertex S. We know that C is somewhere on a line drawn parallel to R A from S, but we do not know exactly where. It is through the triangle corresponding to △ A C E (Fig. 68) that we shall be able to find C. In 440 you learned that △ A C E is a right triangle. How can you find the point corre-

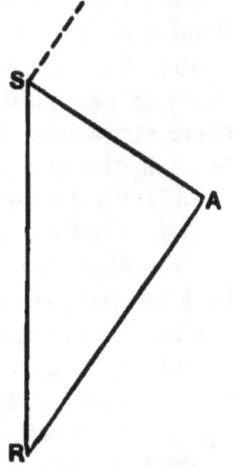

FIG. 69.

FIG. 70.

sponding to E? Review 272, and, with the aid of the principle therein stated, find point C. A C will be the side of the isosceles triangle desired. Draw A B at right angles to A C and equal to it. Complete triangle A B C.

447. Change six different right triangles into isosceles right triangles by the method of 446. Vary the position of the triangles, that you may not depend upon having your figure placed as in Fig. 68.

448. Change a rectangle into a square, following the method suggested by 446. Repeat three times.

449. Change a triangle into a square. Suggestion: First change the triangle into a rectangle.

450. Change a pentagon into a square. Make the steps *in their proper order* perfectly clear.

451. Add a triangle and a quadrilateral, and change the resulting figure into a square.

452. Put three equal squares side by side, so as to form a rectangle. Mould the rectangle thus formed into a square form.

453. Find a square three times as large as a given square. Find a square five times as large as a given square.

454. Cut a square into five equal rectangular strips, finding ⅕ of one side with your dividing tool. Change one of these strips into a square form. How is this new square related to the original square?

455. Find a square ⅕ of a given square. See 454.

456. Find a square ⅖ of a given square. See 454.

457. Make a pentagon. Can you find a square that shall be ⅖ as large as the pentagon?

458. In changing a square into a rectangle, 438, you found that a great many rectangles could be formed equivalent to the square. The shape of the rectangle depends upon the distance which the vertex of the square is moved along C E (see Fig. 68). In changing a rectangle into a square, how many results can you get?

459. Make a square, A B C D, Fig. 71. Change the square into a rectangle eight times by moving each vertex

in the two directions indicated by the arrows at the vertex (Fig. 71).

Move each vertex *the same distance,* and compare the resulting rectangles.

Note.—The object of this exercise is to give you practice in changing squares into rectangles with the squares in different positions. If you find any one position hard for you, practise that case until it is as simple as any of the others.

460. Draw a square A B C D (Fig. 72), and change it, by 438, to rectangle K H A F. The fact that the square is equivalent to the rectangle may be expressed as follows: $\overline{AD}^2 \Leftrightarrow AF \times AH$.

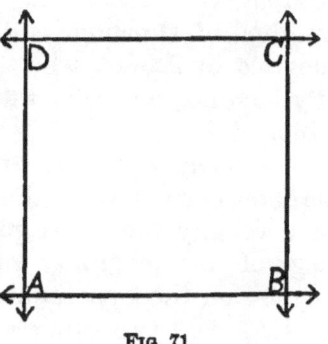

Fig. 71.

Show that $\overline{ED}^2 \Leftrightarrow EF \times ES \Leftrightarrow \square\, SKFE$.

Suggestion: Imagine that D has been moved to A, parallel to N E.

461. What kind of quadrilateral is S H A E, which is made up of the two rectangles S K F E and K H A F?

Show that $\overline{AE}^2 \Leftrightarrow \overline{AD}^2 + \overline{DE}^2$.

462. Does the truth stated in 461 apply to all right triangles?

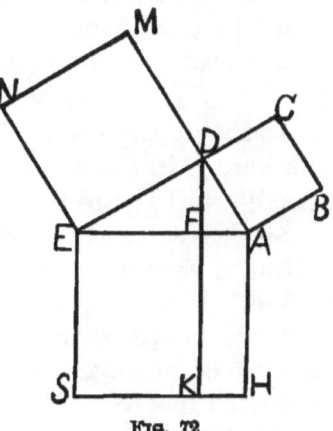

Fig. 72.

463. The principle of 460 and 461 stated in words is as follows: **The square formed on the hypotenuse of a right triangle is equivalent to the sum of the squares formed on its sides.**

This principle is called the **Pythagorean** principle, because the discovery of it is attributed to Pythagoras, a famous Greek mathematician, who lived about 600 B.C., and who did much to encourage the study of Geometry. How Pythagoras made his discovery is not known. The method of showing its truth here given is essentially that devised by **Euclid**, who lived three hundred years later than Pythagoras, and who was the most famous of the ancient Geometers.

464. The Pythagorean principle makes the addition of squares even simpler than the addition of triangles; for it is necessary merely to place the squares to be added on the legs of an imaginary right triangle, as in Fig. 72. The square on the hypotenuse is the square desired.

465. Add two squares. Repeat the process four times, using different squares.

466. Can you form two unequal squares on one line, A B, as a side? If A B is a given line, can you form $\overline{AB}^2$ without any further description?

What is meant by the statement that a square is determined by one of its sides?

467. Can you find the side of a square equivalent to $\overline{MN}^2 + \overline{ST}^2$, where M N and S T are given lines, without constructing any square? Describe your process carefully. It is better, in the majority of cases, not to draw the squares actually, but to use the sides of the squares.

468. Given two lines, a and b, find $a^2 + b^2$ in a square form. Find a square equivalent to $a^2 + a^2$ or $2\,a^2$; to $a^2 + 2\,a^2$ or $3\,a^2$. Find a square equivalent to $3\,a^2$ by 453, and compare the result with that just obtained. Compare your method of getting the side of $2a^2$ here with the method suggested in 315.

469. Given three lines a, b, c. Can you find a square equivalent to $a^2 + b^2 + c^2$? Find $a^2 + a^2 + a^2$ by this method. Find $3a^2$ by the methods of 453 and 468, and compare the results with that obtained here.

470. Change two triangles into squares, and add the resulting squares. Add the two triangles without changing

them to squares; change the resulting triangle to a square, and compare the results obtained by the two methods.

471. Make a square on a given line, a. Make a square on a line equal to $2a$. Compare the two squares, a^2 and $(2a)^2$. How much larger than the first square is the second?

472. Make a square on the line $3a$ as a base, and compare the resulting square with a^2.

473. What is the side of a square equivalent to $16\ a^2$? to $25\ a^2$? to $100\ a^2$?

474. Make a square on the line a. Draw a line M N equal to $3a$ and, at right angles to M N, draw N P equal to a. Join M P, and call it b. How many times as big as a^2 is b^2?

475. Can you devise a simple method of finding $5\ a^2$ from the experiment in 474? Find a square five times as large as a^2 by the method of 453, and compare the result with that obtained by your new method.

476. Can you draw a right triangle by first drawing the hypotenuse? Recall what you know about the middle point of the hypotenuse.

477. Make a line A B, and draw a semi-circumference, of which A B is the diameter. Join any point, C, of this semi-circumference with A and B. Can you tell the size of $\angle$ B C A? Give a reason for your answer.

478. Make a square, a^2, of the same size as that used in 474 and 475. With a line A B equal to $3a$ as a diameter, describe a semi-circumference. See Fig. 73. Make a line A C, called a **chord** of the circumference, equal to a. Draw the chord C B.

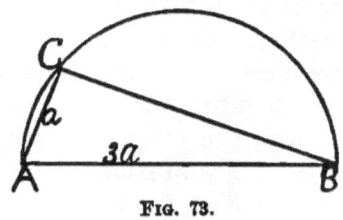

FIG. 73.

Draw squares on A B, A C, and B C.
How much larger than $\overline{AC}^2$ is $\overline{AB}^2$?
How much larger than $\overline{AC}^2$ is $\overline{CB}^2$? Why?

479. Draw a new semi-circumference of the same size as

that of 478, and this time draw a chord A D equal to $2a$. Join D B, draw the squares as before, and compare all the squares with a^2.

480. Draw two squares, a^2 and b^2. Can you find a square equivalent to the difference between a^2 and b^2? Describe the steps carefully.

481. Cut from a square piece of paper, at any place, a smaller square. If the remaining piece of paper could be moulded into a square form, how long would the sides of the square be? Show, by a neat construction, how you would obtain the desired length.

482. Make three pairs of squares and subtract the smaller square of each pair from the larger, obtaining your result in the form of a square.

483. It is a simple matter, as you have seen, to find a square equal to $4a^2$, $9a^2$, $16a^2$, $25a^2$, etc., when a^2 is given. Can you, by simple addition or subtraction of these easily obtained squares, find squares equal to $7a^2$, $13a^2$, $12a^2$, $20a^2$, $3a^2$?

Whenever you see two combinations that will give the same result, it is well to try both and to compare the answers. After the first two or three problems, it is not necessary to draw the actual squares. You will find that you need to deal with the sides only; for a square is easily formed in imagination, if its side is clearly seen.

484. Not receiving sufficient light from a certain square window, a man decided to make an opening twice as large without changing its shape; he, accordingly, made his window twice as long and twice as wide as before. How much more light was let in? What should he have done to find the proper length for his window?

485. Two farmers have square fields to fence; if the field of the first farmer is four times as large as that of the second, how much more fencing will he need? Draw plans of the fields before making a final answer? Can you account for the difference between the relative cost of fencing and the relative size of the fields?

486. On a piece of cardboard draw a diagram, showing how you would cut the cardboard to make it fold into a hollow cube with one-inch edges. If you were to make a hollow cube with two-inch edges, how much more surface would your diagram cover? How much more sand would your second cube hold than the first would hold? How much cardboard would you save in making one large cube like the second one, in place of making enough small cubes like the first to hold an equal amount with the second?

487. If you want to make a cubical box that will have an outside surface just twice as large as that of the first cubical box of 486, how will you proceed to draw the diagram by which to cut out the box?

488. Can you draw a diagram that will fold into a cube with a surface three times as large as that of the first cube of 486?

489. You cannot become too familiar with the peculiarities of squares that have been dwelt upon in this section. You will find later that a great many other figures behave like squares; but this will do you little good, unless you know well the peculiar characteristics of squares, which enable you to add them, to subtract them, to show what effect the change in a side will produce upon the surface, and to show how a change in the size of the square affects the side. You ought also to be expert in moulding any polygon into a square form.

490. Cut from a piece of paper a square 3 in. by 3 in. Can you, by adding strips to two sides only, keep the paper in the form of a square?

If you add strips an inch wide, how long must the strips be to keep the square perfect? How many square inches in the strips added?

491. Cut a strip $\frac{1}{2}$ in. wide of the right length to make a perfect square when added to two sides of the 3 in. square of 490. How many square inches in the strip?

If the strip is 2 in. wide, how long should it be?

492. Imagine a square 7 ft. by 7 ft. How long a strip will be needed to enlarge the square to a square 9 ft. by 9 ft., if the strip is added on two sides? How many square feet in the strip?

493. Imagine a square 80 ft. by 80 ft. How long a strip will be needed to enlarge this to a square 87 ft. by 87 ft. ? How many square feet in the strip?

494. Calculate the lengths and areas of the strips needed in each of the following cases :—square 70 ft. by 70 ft., strip 9 ft. wide; square 130 ft. by 130 ft., strip 5 ft. wide; square 90 ft. by 90 ft., strip 6 ft. wide; square 700 ft. by 700 ft., strip 20 ft. wide.

495. Can you state the law by which you can find the length of the strip from the length of the square and the width of the strip?

496. Make a square 3 in. by 3 in. If you have sixteen square inches of paper which can be cut into rectangular strips, how wide will you make your strips, that you may have a perfect square when you shall have added the strips to the 3 in. square? The strips in this and the following problems are to be added on only two sides of the square.

497. Solve the problem of 496, if the original square is 9 ft. by 9 ft. and the material for addition has a surface of 63 sq. ft.

498. By adding a strip that contained 889 sq. ft. to a square 60 ft. by 60 ft., a square was obtained; how wide and how long was the strip?

499. Calculate the length and width of the strips in the following cases :—square 80 ft. by 80 ft., area of strip 656 sq. ft.; square 50 ft. by 50 ft., area of strip 981 sq. ft.; square 170 ft. by 170 ft., area of strip 2076 sq. ft.

500. Can you make a rule for finding the length and width of a strip needed to make a perfect square, when added to a known square, from the side of the known square and the area of the strip?

501. Arrange 784 sq. ft. in the form of a perfect square

by first making as large a square as you can, using only round numbers like 10, 20, 50, 100, 300, etc., for the side of the square, and by then adding the material left in the form of a strip according to the examples given in articles 496–499, and according to the rule in 500.

The following model for the arrangement of the work is suggested. In the model the problem is to arrange 4489 sq. ft. in a perfect square.

Dimensions of strip.	Material.	Side of square.
	4489	60 (1st guess)
0	3600	
127	889	7 (width of strip)
7	889	
	0	67 answer.

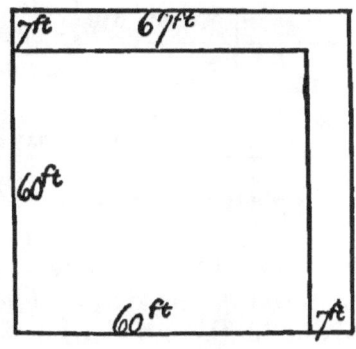

Fig. 74.

Explanation. — A square 70 ft. by 70 ft. would require 4900 sq. ft. of surface; therefore, in round numbers, a square 60 ft. by 60 ft. is the largest possible square. The material left after forming this square (Fig. 74), is 889 sq. ft.; the length of the strip to be added must be at least 2 × 60 ft. or 120 ft.; the width is roughly 889 ÷ 120 or 7; since the strip must be as much longer than 120 ft. as it is wide, 7 ft. is added to 120 ft. to obtain the true length of the strip. The surface covered by the strip is 127 ft. × 7 ft. or 889 sq. ft. All the material is thus used, and a perfect square is formed. In writing the first estimate, 120 ft., of the length of the strip, the cipher was written a little above the line, to leave room for the addition of the width to the length as soon as the width could be guessed. It is well to draw a figure like Fig. 74 to aid the imagination, until the process is perfectly familiar to you, so that no aid is needed.

502. Arrange the following surfaces in square forms, following the model given in 501: 1156 sq. ft.; 6889 sq. ft.;

5184 sq. ft. ; 2401 sq. in. ; 169 sq. yds. ; 729 sq. ft. ; 56169 sq. in. In doing the last problem follow this model :

Dimensions of strip.	Material.	Side of square.
	418609	600 (1st guess)
	360000	
00		
1240	58609	40 { width of 1st strip
40	49000	640 { side of 2d square
0		
1287	9009	7 { width of 2d strip
7	9009	
		647 answer.

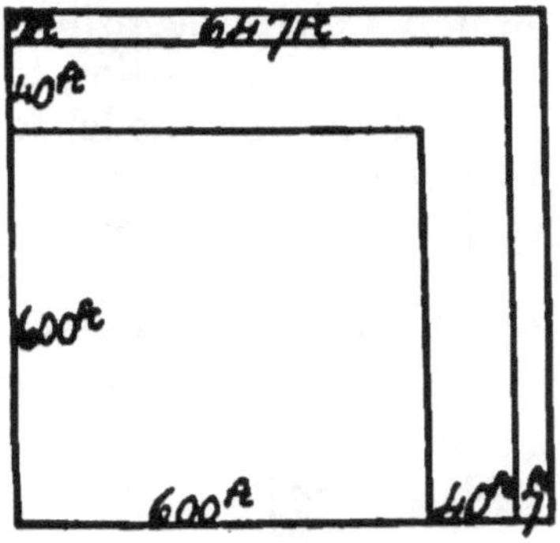

Fig. 75.

In getting the width of the first strip, do not try to guess anything except round numbers like 30, 40, 50, etc. ; and do not try to add all of the material in one strip. The explanation is the same as that given in the previous article ; two steps instead of one are necessary.

503. Arrange the following surfaces in a square form, obtaining the length of the side of the square in each case : 762129 sq. ft. ; 145161 sq. ft. ; 93636 sq. ft. ; 60516 sq. ft. ; 120409 sq. ft.

504. In arranging the following surfaces in a square form, you will find that some material will be left over. Calculate the side of the square and the amount left over in each case :—536 sq. ft. ; 4376 sq. ft. ; 59785 sq. ft.

505. If, after adding one or two strips to a square, there is not enough material left to make a strip 1 ft. wide or 1 in. wide, according to the unit used, it is possible to add strips that are so many tenths of a foot, or tenths of an inch in width ; and then to add other strips whose width is measured in hundredths of a foot or hundredths of an inch, and so on until the material left is not worth noticing. See the following example, where 281 sq. ft. are to be arranged in the form of a square :

CONSTRUCTIONAL GEOMETRY

Dimensions of strip.	Material.	Side of square.	
	281.	10	first guess
	100.		
0			
26	181.	6	width of 1st strip
6	156.	16.	side of 2d square
32.7	25.00		
.7	22.89	.7	width of 2d strip
33.46	2.1100	16.7	side of 3d square
.06	2.0076	.06	width of 3d strip
33.523	.102400	16.76	side of 4th square
.003	.100569	.003	width of 4th strip
	.001831	16.763	answer.

281 sq. ft., therefore, when moulded into the form of a square, will make a square whose side is 16.763 ft.; the part unused, a little under two-thousandths of a square foot, is not worthy of notice in ordinary cases. Multiply 16.763 by 16.763, and see how much the product lacks of amounting to 281.

506. Arrange in the form of a square the following surfaces; calculate the side to the thousandth of your unit:—
69 sq. ft.; 351 sq. ft.; 48 sq. ft.; 10 sq. in.; 5 sq. in.; 3 sq. in.; 2 sq. in.

507. Make a square decimeter. Draw the diagonal and measure it carefully. Arrange in the form of a square two square decimeters, calculating the length of the side as in 506. Compare your two answers.

508. The side or base of a square is sometimes called the **root** of the square. The **square root** of a surface is one of the two equal lines that, multiplied together, will produce the surface; in other words, it is the side of the square into which the surface may be moulded. Thus the square root of 4489 sq. ft. in 501 was found to be 67 ft. This fact would be written in sign language as follows:—
$\sqrt{4489}$ sq. ft. = 67 ft. Notice the sign for square root. You have already found the square root of 1156 sq. ft.,

2401 sq. in., 169 sq. yds., in 502; write the results obtained, making use of the proper signs.

In arithmetic the square root of a number is one of the two equal factors that, multiplied together, will produce the number. The process of getting the square root of a number is the same as that of getting the square root of a surface explained in articles 501–505.

509. In 507 you drew a line equal to $\sqrt{2 \text{ sq. dm.}}$ without any arithmetical work. Can you think of a way to draw a line equal to $\sqrt{5 \text{ sq. in.}}$? Compare your answer with the answer found in 506 by the former method.

510. Draw lines equal to $\sqrt{2 \text{ sq. in.}}$, $\sqrt{10 \text{ sq. in.}}$, $\sqrt{3 \text{ sq. in.}}$; explain your steps; measure the resulting lines; and compare the answers with those obtained in 506.

511. Can you draw a line that shall represent the $\sqrt{\frac{1}{2} \text{ sq. in.}}$; $\sqrt{\frac{1}{4} \text{ sq. in.}}$; $\sqrt{\frac{1}{8} \text{ sq. in.}}$. (See 454.) In actually doing this work, it is better to draw to scale, taking a line much longer than an inch to represent the side of the square inch.

512. On a line mark off lengths equal to $\sqrt{1 \text{ sq. dm.}}$, $\sqrt{2 \text{ sq. dm.}}$, $\sqrt{3 \text{ sq. dm.}}$, etc., up to $\sqrt{10 \text{ sq. dm.}}$

513. On a line mark off lengths equal to $\sqrt{\frac{1}{2} \text{ sq. dm.}}$, $\sqrt{\frac{1}{4} \text{ sq. dm.}}$, $\sqrt{\frac{1}{8} \text{ sq. dm.}}$, etc., up to $\sqrt{\frac{1}{16} \text{ sq. dm.}}$

514. With the scales you have just made in 512 and 513, and with the aid of what you know about adding and subtracting squares, you can find the square root of a great many surfaces with great rapidity. Use the scales in finding the following square roots: $\sqrt{17 \text{ sq. dm.}}$; $\sqrt{12 \text{ sq. dm.}}$; $\sqrt{24 \text{ sq. dm.}}$; $\sqrt{5\frac{1}{4} \text{ sq. dm.}}$; $\sqrt{2\frac{1}{4} \text{ sq. dm.}}$; $\sqrt{\frac{1}{16} \text{ sq. dm.}}$

515. In articles 501–505 you saw how to find the side of a square from its surface. The ability to do this adds greatly to the practical value of the Pythagorean principle that the square on the hypotenuse of a right triangle contains as much surface as the squares on the two legs. If the legs of a right triangle are 7 in. and 3 in., the square on the hypotenuse must contain 49 sq. in. + 9 sq.

in. or 58 sq. in. Show that the length of the hypotenuse is 7.61 in.

516. Measure the length and width of the bottom of a chalk box, and calculate the length of the diagonal across the bottom.

517. Measure the length and width of the school-room, and calculate how far apart the diagonally opposite corners are.

518. A table is 8 dm. long and 6.6 dm. wide; how long a straight stick could be placed on the table so that no point of the stick would be beyond the edge of the table?

519. A drawer is 8 dm. long, 3.5 dm. wide, and 1 dm. deep; what is the longest straight stick that can be placed wholly within the drawer?

520. A ladder 13 ft. long was placed against a tree, with its foot 5 ft. from the base of the tree; how far up the tree did the ladder reach?

521. A boy walked round two sides of a rectangular field, 11 yds. by 60 yds.; how much would he have saved, if he had gone diagonally across the field?

522. In raising a pole 65 dm. long a man rested the upper end against a house; he found the lower end to be 33 dm. from the house; the next time that he rested the pole against the house he found that its end was 16 dm. from the house; how many decimeters had he raised the upper end the second time? Make an estimate of the amount; then calculate it; and compare the results.

523. An isosceles triangle has a base 24 in., and legs 37 in. long; how long is the altitude of the triangle? How many square inches in its surface?

524. An equilateral triangle has sides 12 in. long. Can you calculate its altitude and its area?

525. Make a right angled triangle with one leg 3 in. and the other leg 4 in. Calculate the hypotenuse.

526. The triangle that you have just made is a most useful one in practical work. If you wish to nail two straight sticks together so that their edges shall be at right angles

to each other, you can mark the points 3 in. or 3 ft. along one stick and 4 in. or 4 ft. along the other, and then nail a stick across the two so that there shall be a length of 5 in. or 5 ft. between the points marked. See whether the edges of a box are really at right angles to each other by this test. Show how you could set a corner post at right angles to a platform by the aid of the 3, 4, 5, right triangle.

527. A pyramid is formed by four equilateral triangles resting on the sides of a square base 6 in. by 6 in. Calculate the entire surface of the pyramid.

528. A rhombus has sides 13 in. long and one diagonal 24 in. long; how long is the other diagonal, and what is the area of the rhombus?

529. Divide a circumference whose radius is 6 in. into arcs of 60° each; join the ends of these arcs, forming a regular hexagon. Divide the hexagon into six triangles from the centre, and calculate the area of the hexagon.

530. The end of a house is 20 ft. by 20 ft. to the eaves; the ridge pole is 30 ft. above the ground. Find the length on the roof from ridge pole to eaves, and the number of square feet in the gable end. Can you tell the pitch of the roof?

531. A diamond-shaped window pane has one edge 8 in. long and one angle 60°. Can you calculate the number of square inches of glass in the pane? Would more or less light come in, if the pane were in the shape of a square with one side 8 in.?

An equilateral triangle has sides 18 in. long. Draw two of its altitudes, and calculate how far from the vertices of the triangle they cross each other.

Divide a circumference whose radius is 10 in. into arcs of 90° each; join the ends of the arcs, forming a square. Calculate the sides and the area of this square.

A circumference is drawn through the corners of a square whose sides are 8 in. long. Can you calculate the radius of the circumference?

The diagonals of a rhombus are 12 ft. and 20 ft. respectively. Calculate the sides of the rhombus.

CONSTRUCTIONAL GEOMETRY 103

532. Models of solids having four, eight, or twenty faces in the shape of equilateral triangles, may be made from cardboard by drawing the patterns of Fig. 76, and by cutting the cardboard half through along the lines that separate the triangles. Bend the cardboard until the edges come together, and glue the edges. Calculate the total surface of each solid, if one side of the equilateral triangle is 1 in.

FIG. 76.

SECTION XVII. AREAS OF SIMILAR FIGURES.

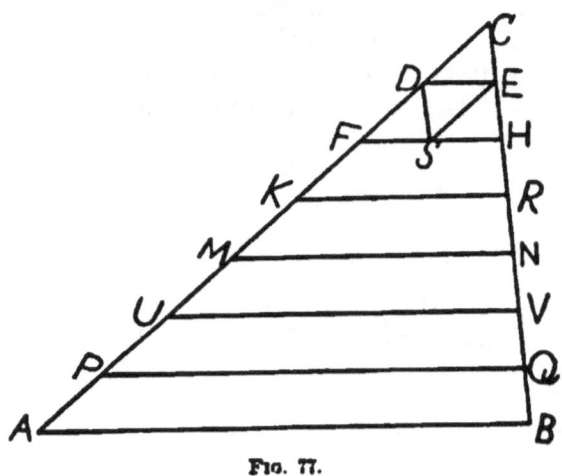

Fig. 77.

533. With your dividing tool, divide one side A C of a scalene triangle, A B C (Fig. 77), into seven equal parts, and draw lines parallel to the base through the points of division. Draw D S parallel to C B, and draw E S the diagonal of the parallelogram D E H S. Show that the trapezoid F H E D is made up of three triangles equal to △ D E C. What part of △ C F H is △ D C E? How much larger is △ F H C than △ D E C? How much longer is F H than D E? F C than D C? H C than E C? Into how many triangles like D E C can you divide the second trapezoid K R H F? Give the number of triangles like D E C in the third trapezoid M N R K; in the fourth trapezoid U V N M; in the fifth trapezoid P Q V U; in the sixth trapezoid A B Q P. How much bigger than △ D E C is △ A B C? How much longer are the sides of △ A B C than the corresponding sides of △ D E C? Compare the sides of △ C K R with the corresponding sides of △ C D E; also compare the surfaces of the two triangles.

Compare the sides of △ C M N with the corresponding sides of △ C D E ; also compare the surfaces of the two triangles. Compare the surfaces and the corresponding lines of △ C U V and C F H ; of △ C P Q and C M N ; of △ C A B and C U V.

534. Triangles like those studied in 533, where the corresponding angles are equal, and where the corresponding sides have the same relation to each other, are called **similar** triangles. The investigations suggested in 533 point to the law that similar triangles have the same peculiarities as squares in the matter of their surfaces. Thus you have learned that, to quadruple a given square, you make a square with its sides twice as long as those of the given square. In the same way, to quadruple a given triangle, you make another triangle similar to it in shape, with the sides respectively double those of the given triangle. Compare the triangle in Fig. 77, that is nine times △ C D E with △ C D E, and see if its sides have the relation to those of △ C D E that you would expect from your knowledge of squares. Do the same with the triangle that is twenty-five times as large as △ C D E ; with the triangle that is thirty-six times as large as △ C D E.

535. In Fig. 77, one of the triangles is two and one-quarter times as large as △ C F H. Find this triangle and record the result in this way :—△ .... ⌒= $\frac{9}{4}$ △ C F H. Do the sides of the triangle found bear the relation to those of △ C F H that you would expect from your knowledge of squares ?

536. The sides of △ C U V are $\frac{5}{6}$ of the corresponding sides of △ C F H. If similar triangles resemble squares in the matter of areas, how much larger than △ C F H should △ C U V be ? Test your answer by counting the number of triangles like △ C D E in each of the larger triangles concerned.

537. All of the experiments made from 533 to 536 confirm the statement that **similar triangles are like squares in the way in which the areas change as the sides**

change. To give a complete proof of this statement would carry us too deeply into the principles of Geometry for our present purposes. The principle may be accepted as a good working principle until it can be shown to fail, or until it can be proved to be true in all cases.

538. A little further study of Fig. 77 would show that the altitudes of the various similar triangles are related to each other just as are the sides of the corresponding triangles; thus the sides of △ C F H are $\frac{3}{7}$ of the corresponding sides of △ C U V, and the altitude of △ C F H is also $\frac{3}{7}$ of the altitude of △ C U V, because the parallels D E, F H, and K R, etc., cut all lines between C and A B into seven equal parts. Draw a figure like Fig. 77, and draw the altitude from C to A B, and show that it is divided into seven equal parts by the parallels. This fact gives additional reason for believing the principle of 537 to be true; for the essential peculiarity of squares, so far as their area is concerned, is that no change can take place in the base of a square, unless it also takes place in the side or altitude of the square; if the base is made $\frac{3}{7}$ as large, the altitude must also be made $\frac{3}{7}$ as large, or the figure ceases to be a square; the area of the square, which is nothing but the product of the base and altitude, is affected by both changes, and therefore becomes $\frac{3}{7}$ of $\frac{3}{7}$ as large, or $\frac{9}{49}$ as large. Now this same peculiarity is shared by similar triangles; if the base is made $\frac{3}{7}$ as large, the altitude must also be made $\frac{3}{7}$ as large, or else the new triangle ceases to be similar to the old; the area, which is merely the product of the base and the half-altitude, is affected by both changes, and becomes $\frac{3}{7}$ of $\frac{3}{7}$ as large or $\frac{9}{49}$ as large.

539. If similar triangles are like squares, they can be added like squares. Suppose you wish to add △ C K R to △ C M N (Fig. 77); place C K at right angles to C M (Fig. 78), as if you were to add $\overline{CK}^2$ to $\overline{CM}^2$; K M will be the side of the new triangle corresponding to C K and C M of the given triangles. Make a triangle with the angles of △ C K R, starting with K M, Fig. 78, in the position of

C K, Fig. 77. The triangle thus formed will be like △ C K R and C M N in shape, and will be equivalent to their sum.

In Fig. 77 there is a triangle that is equivalent to △ C K R + △ C M N. Find this triangle, and compare it with the triangle that you have just constructed.

540. Make three pairs of similar triangles, and add the triangles of each pair by the method described in 539.

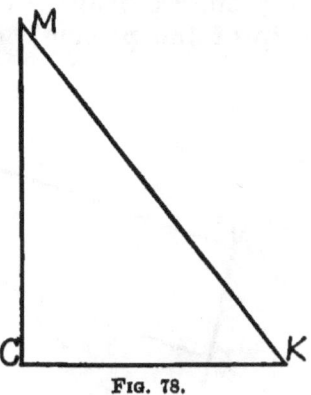

Fig. 78.

541. Can you subtract one triangle from a larger triangle similar to it? Give two illustrations.

542. Can you find a triangle twice as large as another triangle and similar to it? Suggestion: Think how you double a square.

543. Make a triangle A B C; can you find a similar triangle equivalent to $3 \times$ △ A B C? to $4 \times$ △ A B C? to $5 \times$ △ A B C?

544. Given a triangle, A B C, can you find a similar triangle equivalent to ½ of △ A B C? to ⅓ of △ A B C? to ¾ of △ A B C? to ⅘ of △ A B C?

545. Polygons are said to be similar to each other when their corresponding angles are equal, and when their corresponding sides have the same relation to each other.

Make a pentagon, and then, with the aid of your dividing tool and protractor, make another pentagon similar to it and with sides ⅝ as long as those of the first pentagon.

546. A convenient method of making one polygon similar to another is illustrated in Fig. 79.

The problem is to make a pentagon similar to A B C D E with sides twice as long respectively.

Divide A B C D E into three triangles by the diagonals A C, A D. Extend A B to H, making A H = 2 A B. Draw H K ∥ B C meeting A C extended in K. In the same way draw K M | C D and M N | D E, completing the pentagon A H K M N. The corresponding angles of the two polygons are plainly the same. Why? Can you, with the help of the principle of 325, show that H K, K M, M N,

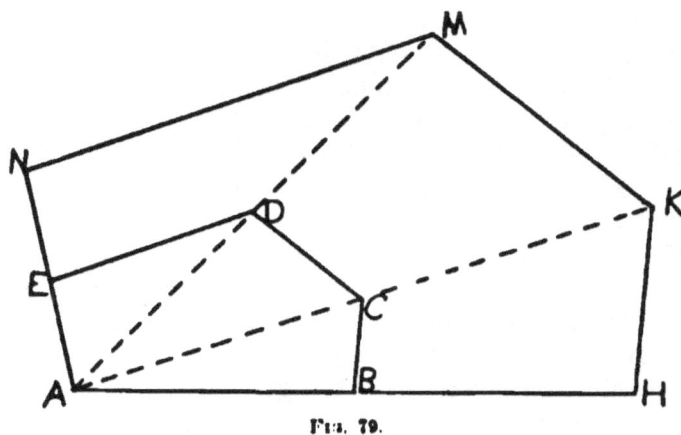

Fig. 79.

and N A are respectively twice as long as the corresponding sides of A B C D E?

How much larger is each triangle of A H K M N than the corresponding triangle of A B C D E? How much larger is A H K M N than A B C D E?

547. Make a hexagon, and then make a similar hexagon with sides ⅓ as long as those of the first, using the method described in 546.

Compare the areas of the corresponding triangles and of the two hexagons.

548. Do similar polygons resemble squares in respect to their surfaces?

549. Can you make a parallelogram twice as large as a given parallelogram, and similar to it?

550. Can you make a pentagon similar to those in Fig. 79, and equivalent to their sum?

551. Can you make a pentagon similar to those in Fig. 79, and equivalent to their difference?

552. Can you make a hexagon $\frac{4}{5}$ of a given hexagon? $\frac{3}{5}$ of a given hexagon? $\frac{1}{2}$ of a given hexagon? In each case the hexagon must be similar to the given hexagon.

553. Can you make a circle equivalent to the sum of two given circles?

554. Can you subtract one circle from another, giving the difference in the form of a circle?

555. Can you change a circular ring, or band, into a circle?

556. Can you make a circle twice as large as another circle? three times as large? four times as large?

557. One circular hole was found to have $\frac{2}{3}$ of the diameter of another; how much smaller than the second hole was the first hole?

558. If you wish to double the amount of light coming through a circular window, should you double the diameter of the window?

559. Draw a circumference, Fig. 80, and with your dividing-tool divide the diameter into five equal parts; draw semicircumferences on the diameters A B, A C, A D, and A E above the diameter A F, and semicircumferences on the diameters B F, C F, D F, and E F below the diameter A F.

What part of the semicircle A H F is A K B?

What part of the whole circle is semicircle A K B?

What part of the whole circle is semicircle B M F?

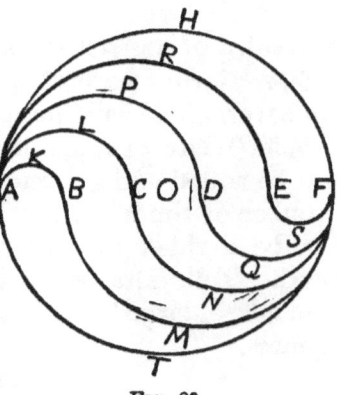

Fig. 80.

The horn-shaped figure A K B F T A is made up of the semicircles A K B + A T F − B M F. What part of the whole circle is the horn?

If the corresponding lines of similar figures have the same relation to each other, what part of the semicircumference A H F should the semicircumference A K B be? What part of the semicircumference A H F should the semicircumference B M F be?

Which is the longer path from A to F, the path by the semicircumference A T F, or that by the two semicircumferences A K B and B M F?

560. By adding and subtracting the right semicircles in Fig. 80, the surface of the strip A L C N F M B K A can be found. Can you select the right semicircles, and can you estimate what part of the whole circle each semicircle is, and what part of the whole circle the strip is? Can you estimate the perimeter of the strip?

561. In 559 you saw that the horn A K B F T A is equivalent to the semicircles $A K B + A T F - B M F$; can you find one semicircle that shall be equivalent to $A K B + A T F - B M F$? See 553 and 554.

562. Make a circle and divide the diameter into eight parts; draw semicircumferences, as in Fig. 80, cutting the circle into eight strips. Try to show that the circle is thus divided into eight equivalent strips.

563. Can you find a circle one-half of a given circle? one-third of a given circle? two-fifths of a given circle?

564. Review carefully Sections XV., XVI., and XVII. (in several lessons if necessary).

565. Write a sketch of the square, describing how squares can be added and subtracted; how they can be enlarged any number of times without losing their shape; and how they can be divided into as many small squares as may be desired. Tell, also, what other figures can be added, subtracted, enlarged, or diminished by the methods used for squares.

SECTION XVIII. CIRCLES AND INSCRIBED ANGLES.

566. A circle is a figure bounded by a line, called the circumference, every point of which is equally distant from a point within, called the centre. The sign for circle is ⊙, plural ⊙s ; and that for circumference is ○, plural ○s. You have already learned that you can measure an angle by placing its vertex at the centre of a circle whose circumference has been divided into three hundred and sixty equal parts, and by noting the number of these parts, or degrees, between the sides of the angle. Any portion of a circumference is called an arc, and the angle formed by the radii through the ends of an arc is said to subtend the arc. Thus in Fig. 81 ∠ B O A subtends the arc B A, written B̂A. The number of degrees in the arc B A is the same as the number of degrees in the angle B O A at the centre; if B̂A = 60°, ∠ B O A = 60° ; if B̂A = 50°, ∠ B O A = 50° ; if B̂A = x°, ∠ B O A = x° ; so that **an angle at the centre of a circle has the same name as the arc which it subtends.** This principle is simply a restatement of the principle of the protractor.

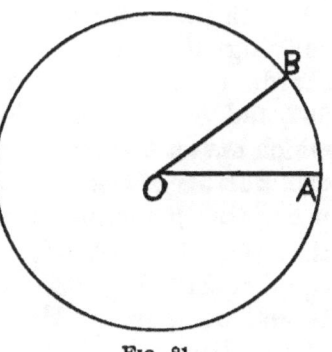

Fig. 81.

567. In using a protractor you have always placed the centre of the instrument upon the vertex of the angle. The peculiarities of circles enable us often to measure angles without a protractor, and also to measure angles whose vertices are not at the centre of the circle. In Fig. 82 A O E is a diameter, and ÂB = 60°. Can you, without a protractor, give the value in degrees of all of the angles in the figure?

Can you see any simple relation between ∠ B E A and ⌒BA?

568. Imagine, or draw, an arc A C = 70°, and the diameter A E. Can you tell how many degrees there are in the angle C E A? Try arcs of 90°, 75°, 30°, 120°, and 150°, from A; and, after joining the end of each arc with E, find the value of each angle formed at E with the line E A. Draw a figure like Fig. 82 for each case, making the arc A B the proper length each time.

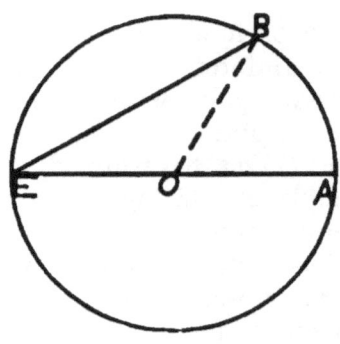

Fig. 82.

569. An angle, placed as angle B E A is placed in Fig. 82, with its vertex on the circumference, and with its sides running to other points of the circumference, forming **chords**, is called an **inscribed angle**. The experiments of 567 and 568 suggest the principle that **inscribed angles, which have a diameter for one side, subtend arcs of twice their own number of degrees**. Try to prove this principle by showing that if ∠ E = $x°$, ⌒AH must be equal to $2x°$ Fig. 83. Draw H O and compare ∠ H O A with ∠ x. How is ∠ H O A related to ⌒HA?

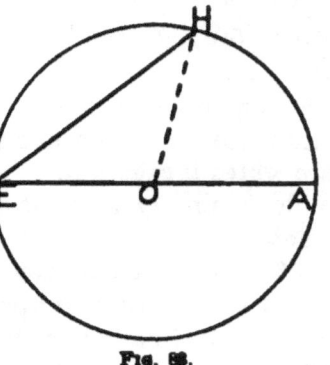

Fig. 83.

570. What is the largest inscribed angle that you can draw with one side a diameter?

From one end of a diameter draw chords that shall make angles of 60°, 45°, 30°, 22½°, and 75° with the diameter, using your compasses and ruler only.

571. Place the inscribed angle x, so that the centre of the circle shall come between the two chords which form its sides, as in Fig. 84. Will $\stackrel{\frown}{AH}$ still be equal to $2x°$? Draw the diameter E O K, dividing $\angle x$ into two pieces, each of which has a diameter for one side. If you give to piece A E K the name y, the piece K E H will be $x - y$. $\stackrel{\frown}{AK} = 2y°$; $\stackrel{\frown}{KH} = 2x° - 2y°$; $\therefore \stackrel{\frown}{HA} = 2y° + 2x° - 2y° = 2x°$, so that $\stackrel{\frown}{HA} = 2x°$ as before.

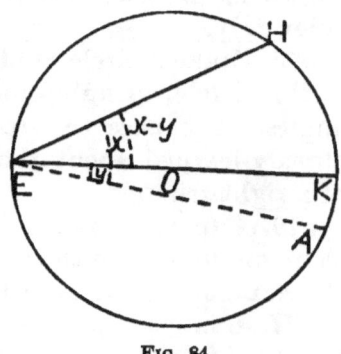

Fig. 84.

If $\angle x$ be inscribed, as in Fig. 85, in such a way that both chords come on the same side of the centre, will the arc A H still contain twice as many degrees as the angle x? Draw E O K, the diameter, and call $\angle$ A E K, $\angle y$, as before; $\angle$ H E K $= \angle x + \angle y$. $\stackrel{\frown}{KA} = 2y°$, and $\stackrel{\frown}{HK} = 2x° + 2y°$. $\therefore \stackrel{\frown}{HA} = \stackrel{\frown}{HK} - \stackrel{\frown}{KA} = 2x° + 2y° - 2y° = 2x°$ as before. We therefore can say that **any inscribed angle subtends an arc of twice its own number of degrees.**

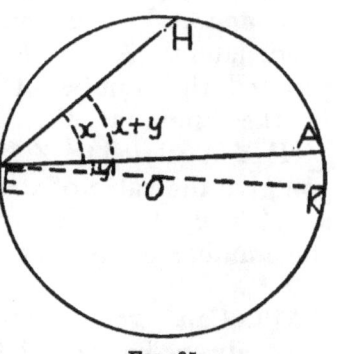

Fig. 85.

572. How large an arc must an inscribed angle of 90° subtend? an angle of $22\frac{1}{2}°$? an angle of 150°?

573. Several arcs of a circumference, when measured, were found to be respectively, 70°, 83°, 90°, 340°, and 180°; how large was each inscribed angle that subtended these arcs?

574. An inscribed angle is said **to stand on** the arc which it subtends. Can you say why all inscribed angles in one circle which stand on the same arc must be equal? Make a circle, and inscribe any angle A E H. Make six other angles in the same circle equal to ∠ A E H, using your ruler only.

575. Make a circle, and make six right angles by means of this circle, using no tool except your ruler to make the angles. Compare the results here with what you have already learned about the middle point of the hypotenuse of a right triangle.

576. Can you inscribe a regular hexagon in a circle? Can you determine the number of degrees in each angle of the hexagon by the arc subtended?

577. Can you inscribe a regular octagon and find the number of degrees in each angle?

578. Inscribe an angle of 75° (∠ A E H, Fig. 86). Choose any point V in the arc A H and draw the chords H V and A V completing the quadrilateral A V H E. Can you tell the number of degrees in the angle at V? How is ∠ H V A related to ∠ E? Can you give the value of the angles at H and A? Can you tell the number of degrees in their sum?

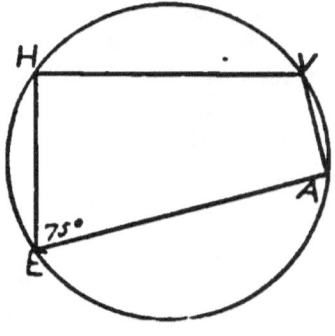

Fig. 86.

579. Can you show what must always be the relation between the opposite angles of an inscribed quadrilateral? Is there any parallelogram whose opposite angles have this required relation?

580. Can you make a circumference pass through the four corners of a rhomboid? of a rhombus? of a rectangle? of a square?

581. Draw a circle, Fig. 87, and then draw two parallel chords, A E and H K. By the aid of A K can you show that the arc A H is equal to the arc K E, and thus establish the principle that **parallel chords intercept equal arcs?**

582. Any chord divides a circle into two **segments**; for example, A E, Fig. 87, divides the circle into the two segments A H K E, and A V E. Segments are named by the number of degrees in their arcs; thus a segment of 200° is one whose arc is 200°; a semicircle is a segment of 180°. If one of the segments into which a chord divides a circle is 120°, what is the other segment? An angle is inscribed in a segment when its vertex is in the arc of the segment, and when the sides of the angle pass through the ends of the chord that makes the segment. Thus the angle A F E would be an angle inscribed in the segment A V E, Fig. 87.

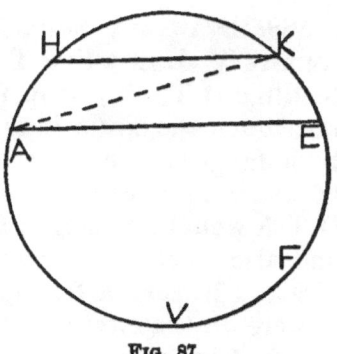

Fig. 87.

583. An angle A T E is inscribed in a segment of 100°; what is the value of the arc A E on which the angle stands? of the arc A T E? of the angle A T E?

584. How large a segment is required that an angle of 120° may be inscribed in it?

585. What can you say of all angles inscribed in the same segment? Why?

586. If an angle is inscribed in a semicircle, what must it be?

587. Of two inscribed angles, which is inscribed in the larger segment, the larger angle or the smaller one?

588. If a line merely *touches* a circle at one point, so that, when extended in either direction, it does not pass within, or out, the circle, it is called a **tangent** to the circle; the point at which it touches the circle is called the **point of contact.** In Fig. 88, T A is a

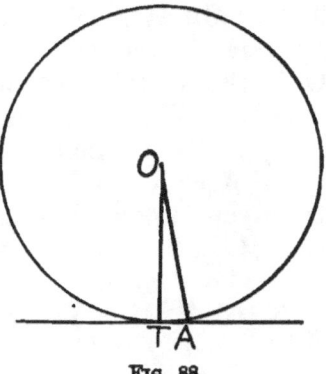

Fig. 88.

tangent to the circle with its point of contact at T. Can you explain why the name *tangent* was selected for this line?

589. Draw any radius, as O T, Fig. 88, and, at its extremity T, draw a line T A ⊥ O T. Do this without extending O T, if you can. Can you show that T A cannot touch the circle in any point except T, so that it must be a tangent? Suggestion: Let A stand for any point of the line except T, and show what kind of triangle O T A would necessarily be, if A were on the circumference, and also what would be the value of the ∠ O A T in such a case. Is such a triangle as △ O T A would become, if A were on the circumference, a possible triangle? Why?

590. Choose several points on the circumference of a circle, and draw tangents touching the circle at these points respectively.

591. The angle A T H, Fig. 89, made by a tangent T H and a chord T A through the point of contact is not called an inscribed angle, but like an inscribed angle it contains only one half as many degrees as the arc T E A subtended by the chord T A. Study △ T F O formed by drawing the radius O T and O F the bisector of the angle T O A. What kind of triangle is △ T O A? Select as many pairs of equal angles as you can from your knowledge of the nature of △ T O A and from the method of drawing O F. How many degrees in ∠ F T O + ∠ T O F? How many degrees in ∠ F T O + ∠ F T H? Why must ∠ F T H or ∠ A T H = ∠ T O F? Now ∠ T O F has as many degrees as T̂E, or one half as many degrees as T̂A. Hence ∠ A T H has one half as many degrees as T̂A, or **an angle formed by a tangent and a chord**

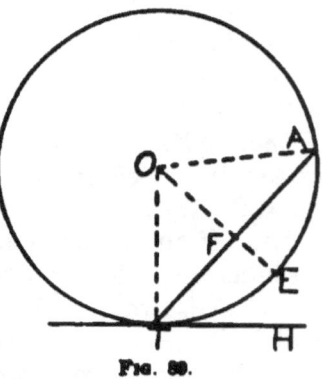

Fig. 89.

through the point of contact contains one half as many degrees as the intercepted arc.

592. How is any angle inscribed in the larger segment formed by T A, Fig. 89, related to ∠ A T H ? How is any angle inscribed in the smaller segment A E T related to ∠ A T H ?

593. A chord A B intercepts an arc of 80°. How many degrees are there in the angle between A B and the tangent at A ? How many degrees in the angle between B A and the tangent at B ?

594. Draw the tangent to the circle of Fig. 89 at A as well as at T. Can you give any reason for thinking that the angle between the tangent at T and T A is the same as the angle between the tangent at A and A T ?

595. What kind of triangle will be formed by a chord and the two tangents at its extremities ?

596. Draw a circle with centre O, Fig. 90, and choose a point, P, outside the circle. It is required to draw a line from P that shall be tangent to the circle.

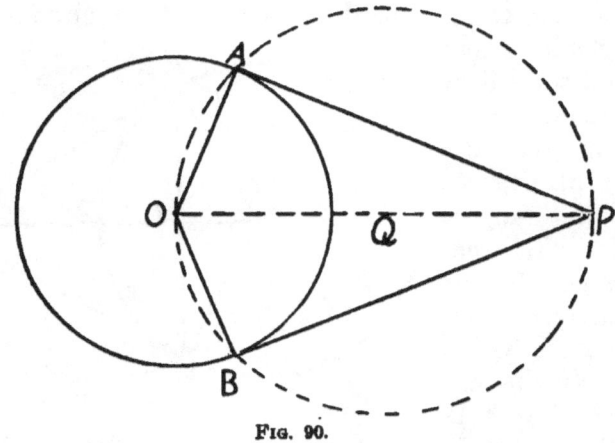

Fig. 90.

If O A P is a right angle, A P will be tangent (589). Study Fig. 90, and try to explain what principles were used to determine the points A and B that ∠s O A P and

O B P might be 90°. Can you see what help line was drawn first?

597. Make four different circles, and draw tangents to them from points selected outside of the circles.

598. If, in Fig. 90, O A is 5 in., and O P is 13 in., how long is A P? What principle enables you to answer this question?

599. A ball, whose radius is 7 in., rests in a hollow cone so that its points of contact with the cone are 24 in. from the apex of the cone; how far is the centre of the ball from the apex?

600. If the radius of the earth is 4000 miles, how far out to sea does the horizon of a person extend, if he is 6 ft. tall, and if he is standing at the water's edge?

601. How far can the light of a light-house 165 ft. high, be seen from the surface of the water?

How far can the same light be seen from a point in the rigging, thirty feet above the surface of the water? Consider the radius of the earth 4000 miles in both cases.

602. Draw a circle with centre at O; then take P at a distance from O equal to the diameter of the circle. Draw the tangents P A and P B, and the chord of contact, A B. Can you give the number of degrees in all the angles formed, with reasons in each case? How many degrees in the arc A Q B? Why? In the arc A O B? Why? If

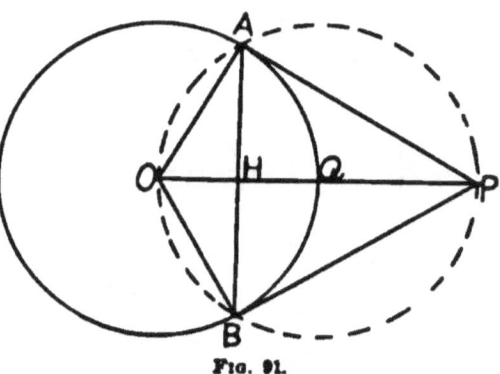

Fig. 91.

the radius of the first circle is 8 in., can you find the lengths of all the lines in the figure, curved and straight?

603. What part of the circumference is AQB, Fig. 91? What part of the earth's surface would a man see were he to

ascend 4000 miles above the earth's surface? How far would he have to go to see one half of the earth's surface?

604. In Fig. 91, if you call angle A P B, $x°$, can you say how many degrees there must be in the arc A Q B?

Two tangents make an angle of 30° with each other; how long an arc do they inclose?

If you want tangents to meet at an angle of 45°, how many degrees apart must you take the points of contact?

605. Can you state a principle concerning the relation between the angle which two tangents form and the arc included between the points of contact?

606. In Fig. 91, assuming that the radius is 8 in., can you find the number of square inches inclosed between the tangents and the arc A Q B? Find also the area of the oval A O B Q; of the smaller segment cut off by A B; and of the triangle A O B.

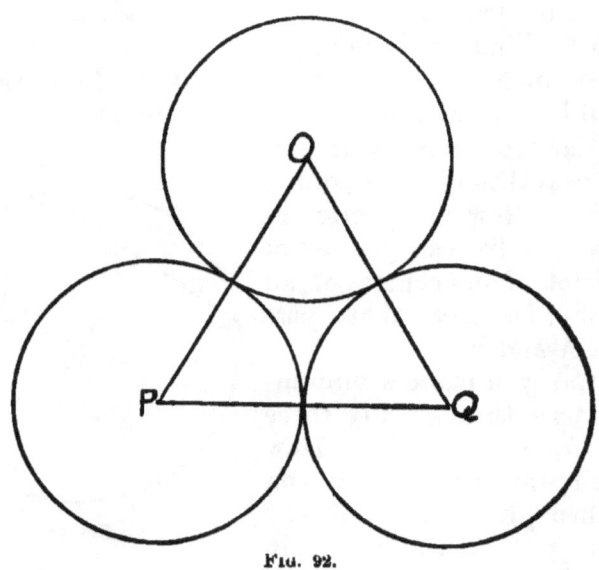

Fig. 92.

607. Three circles of equal size with a radius of 6 in. are placed in contact as in Fig. 92.

Can you find the area of the space between the circles?

What is the area of the parts of the circles without the triangle of centres O P Q?

608. Draw a circle, and draw tangents to this circle at points D, F, and L, Fig. 93, which are 120° apart, thus forming △ A B C. Draw △ D F L, and also △ H P E, whose vertices are at the middle points of the arcs F L, L D, and D F, respectively. Can you account for the shape of the triangles formed? How much smaller than the outside triangle is the smallest triangle? (Recall the principle about the surfaces of similar triangles.) What part of the intermediate triangle is the hexagon? Find the number of square inches in each triangle and in the hexagon, if A B = 6 in. Find also the area of the circle.

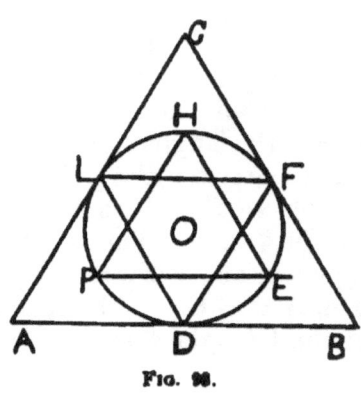

Fig. 93.

609. Can you make a circumference pass through two points, A and B? How many circumferences? What can you say of the position of the centres of all the circumferences that pass through A and B?

610. Can you make a circumference pass through any three points, A, B, and H? How many circumferences can you draw through A, B, and H? Can you put A, B, and H in such positions that a circumference cannot be drawn through them?

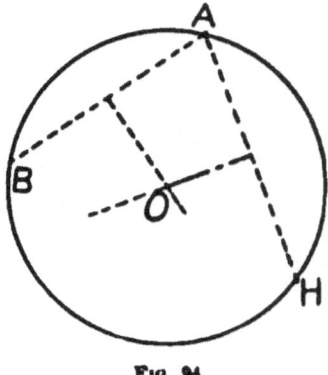

Fig. 94.

611. Under what circumstances can you draw a circumference through four points? Can you recall any quadri-

lateral through whose corners you could draw a circumference?

612. Make a line, and a point without the line; can you draw a circumference that shall merely touch the line, and at the same time pass through the point? How many circumferences fulfilling these conditions can you draw?

613. Can you draw a circumference that shall be tangent to the line of 612 at a chosen point, and at the same time pass through the point without the line?

614. Can you make a circumference with a given radius, touch a given line, and pass through a given point? Can the radius be made too long? Can it be made too short?

615. Can you make a circumference pass through the vertices of a scalene triangle?

616. Using some round object as a guide, draw a circle; can you find the centre of the circle?

617. Draw free hand a curved line; can you tell whether or not the curve is a piece of a circumference?

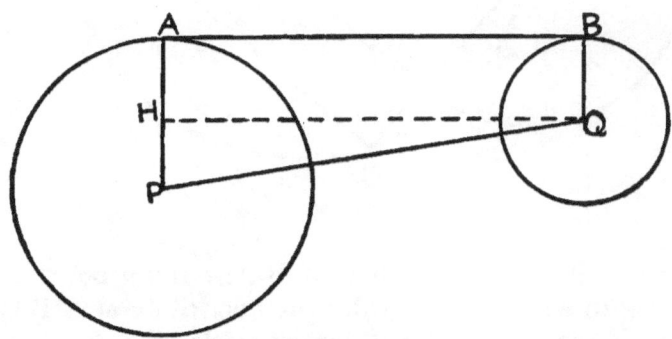

Fig. 95.

618. Can you make a circle touch two intersecting lines? How many answers can you obtain for one pair of intersecting lines?

619. Can you make a circle touch the three sides of a triangle?

620. Can you make a circle touch two lines that do not

meet within the boundaries of the paper, although they are not parallel lines?

621. At two points, A and B, of a line (Fig. 95) draw two circles tangent to the line, using different radii of such lengths that the circles will have no point in common.

Draw the radii, A P and B Q, and join the centres of the circles by the line P Q.

What kind of quadrilateral is A B Q P?

How many of its angles do you know?

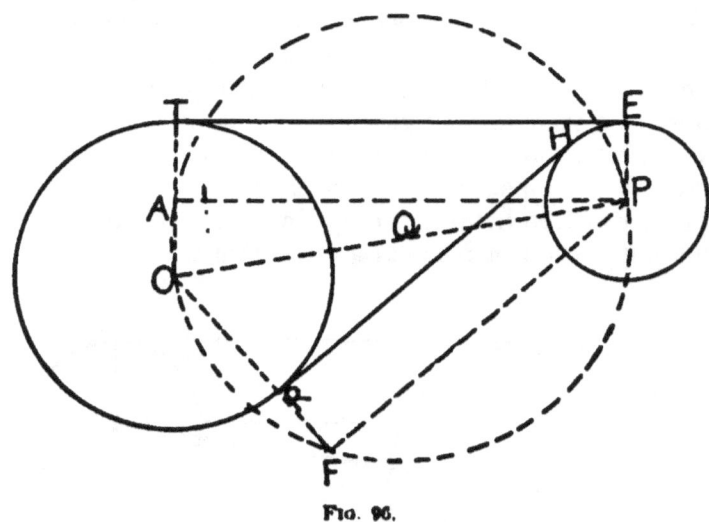

Fig. 96.

Draw a line Q H parallel to B A. Describe the two figures into which Q H divides the quadrilateral A B Q P.

622. If the radius of the larger circle, Fig. 95, is 8 in., and that of the smaller is 3 in., and if the centres of the circles are 13 in. apart, can you find the length of the common tangent A B? of H P? of H Q?

623. Can you make a trapezoid, knowing the two bases and one leg, so that the bases will make right angles with the unknown leg? Suggestion: Try to reproduce Fig. 95 from three lines equal in length to P Q, A P, and B Q.

CONSTRUCTIONAL GEOMETRY

624. Can you draw a line that shall touch two given circles?

Study Fig. 96, where the two given circles have their centres at O and P, and where, by the aid of the circle with O P as diameter, the tangent T E was determined. Compare the quadrilateral O T E P with the quadrilateral P A B Q in Fig. 95.

In Fig. 96 a line, R H, is found, which touches the two circles and passes between them. Can you see how points F and R were determined when this **interior tangent** was drawn?

625. Make two circles and draw an interior tangent to them. Repeat the problem with different circles twice.

SECTION XIX. VOLUMES.

626. In 367 and the following exercises you learned that the product of two lines represents a surface; that the simplest surface that you can form, in multiplying two lines together, is the rectangle; but that this surface can be moulded into many other forms.

627. Multiply a line 7 cm. long by one 4 cm. long; the rectangle, representing their product, contains 28 sq. cm. See rectangle A B C D (Fig. 97), where the 28 sq. cm. are outlined. Draw a line, B F, 5 cm. long, that shall be perpendicular to both B A and B C, and complete the **rectangular parallelopiped** A B C D E F K. If B S is taken equal to one centimeter, and if the figure is divided as represented, what are the dimensions of the solid S B M. What is its name? How many solids like it could be made out of the solid H S T D, which forms the top layer? How many layers like H S T D could be cut from the whole solid? How many cubic centimeters (ccm.) could be cut from the whole solid? How do you obtain the answer to the last question?

628. How many ccm. of wood are there in a rectangular piece 8 cm. long, 5 cm. wide, and 3 cm. thick.

How many cubic inches (cu. in.) of ice could be cut from

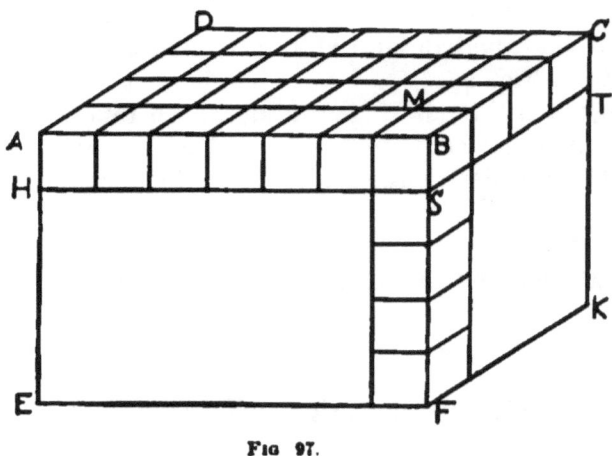

Fig 97.

a rectangular block of ice 24 in. long, 15 in. broad, and 8 in. thick?

629. Just as you saw that the product of two lines represents a surface, you can now see that the product of three lines represents a solid. The simplest solid that you can form, in multiplying three lines together, is the **rectangular oblong**, or rectangular parallelopiped formed by using the three lines for the edges which are at right angles to each other.

The product of two of the lines gives the base of the oblong; the third line is the altitude of the oblong. It is not necessary to form a mental picture of the small cubes into which the oblong may be divided each time that you multiply three lines; in fact it is better to think of the space occupied by the solid as a whole.

630. Make four sets of three lines each, and draw the rectangular oblong represented by the product of the three lines of each set.

631. Draw the oblong of Fig. 97 again, this time without the additional dividing lines. You recall that its dimensions are 7 cm. by 4 cm. by 5 cm. Imagine a piece cut from the solid at the left end along the line V T M, and imagine this piece transferred to the right end, with the face D A E

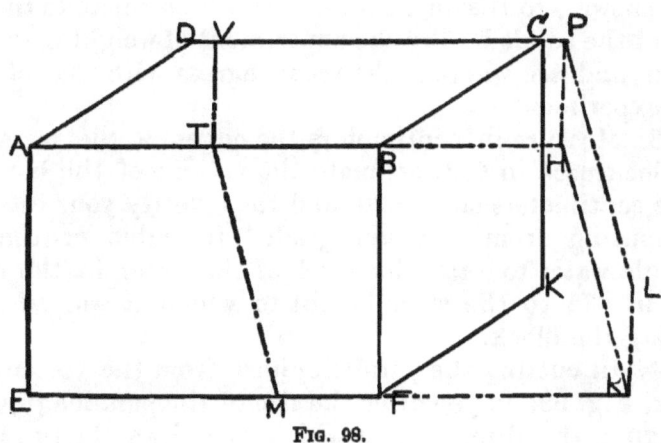

Fig. 98.

placed upon its equal, C B F K. Will the new solid V T M R L P H occupy more or less space than the old oblong? Will its dimensions be changed? (It is supposed that the cutting is done without waste.) What shape will the faces of the new solid have? Will the twelve edges of one solid be any longer than the twelve edges of the other? Why? Which of the new faces, if any, cover any larger surface than the corresponding faces of the oblong? Which solid has the larger upper base? Which lines must you multiply together to obtain the upper base V T H P? What lines must you multiply together to obtain the solid V T M R L P H?

632. The solid V T M R L P H (Fig. 98), is called a **paralellopiped**. Do you understand why? If possible secure two blocks of wood, one a rectangular parallelopiped, and the other a parallelopiped, both of which have been cut from one long rectangular block. The two parallelopipeds should have one edge (corresponding to edge D C, Fig. 98) the same in both. After coating the blocks with paraffine, that they may not absorb water, sink first one and then the other in a vessel of water, noting carefully how high the water rises in the vessel each time. Does the experiment confirm your answers to the questions of 631 which relate to the volume of the solids? Try the experiment of weighing the two blocks, and see whether the result agrees with that of your first experiment.

633. Measure in centimeters the edges of the rectangular block used in 632, estimate the volume of the block in cubic centimeters as in 628, and then verify your estimate by pouring from a vessel graded in cubic centimeters enough water to raise the level of the water in the vessel used in 632 to the same height to which it was raised by sinking the block.

634. In cutting the parallelopiped from the rectangular block, Fig. 98, do you alter the size of the parallelopiped by changing the directions in which you draw the lines V T and T M to guide your cutting? Do you change the area of the upper face as you change the direction of V T? Can you draw V T and T M in such a way that the parallelopiped will have four of its faces rectangles? two of its faces rectangles?

635. Can you multiply three lines together so that the product will be a rectangular parallelopiped? a parallelopiped with four faces rectangles? with two faces rectangles? with no faces rectangles?

636. Can you give any conditions under which two parallelopipeds of different shapes will occupy the same amount of space? Think of the corresponding conditions for equivalent parallelograms.

637. Draw any parallelopiped that is not rectangular, and draw the lines whose product represents the volume of the solid. Form the rectangular parallelopiped that is equivalent to the first solid.

638. Take a block of wood in the form of an oblique-

angled parallelopiped ; find the lines whose product represents the volume of the parallelopiped, measure these lines in centimeters, estimate the volume of the solid, and test your answer by the method of 633.

639. The principle to be derived from Articles 627–638 is that **a parallelopiped is the product of its three dimensions,** or from another point of view, **the volume of a parallelopiped is the product of its base by its altitude.** Just as a rectangle can be moulded into any parallelogram which has the same base and altitude as the rectangle, a rectangular parallelopiped can be moulded into any oblique parallelopiped which has the same base and altitude as the rectangular parallelopiped.

640. Cut a rectangular block of wood into two pieces by sawing along two diagonally opposite edges. The resulting pieces are called **triangular prisms**; the two triangular faces form the bases of the prism, and the three faces in the shape of a parallelogram form the lateral surface. When a rectangular parallelopiped is cut into two triangular prisms, as in this case, the prisms are called **right** prisms; for a right prism is one whose lateral edges are at right angles to the two bases. Do the two right prisms formed in the experiment occupy the same amount of space ? Test your answer to the question by weighing the two prisms, or by sinking them in a vessel of water, and by noting the height to which the water rises in each case.

641. Can you find the three lines whose product represents the volume of one of the prisms of 640 ? Can you draw the rectangular parallelopiped that will occupy the same space as the prism ?

642. Can you show that the product of two of the three lines found in 641 represents the base of the prism, and that the third line is the lateral edge, and at the same time the altitude of the prism ?

In doing this you illustrate an important principle of Geometry : **The volume of a right triangular prism is the product of its base by its altitude.**

643. Imagine the oblique parallelopiped (A B C D E F H, Fig. 99), none of the faces of which are rectangles, to be cut

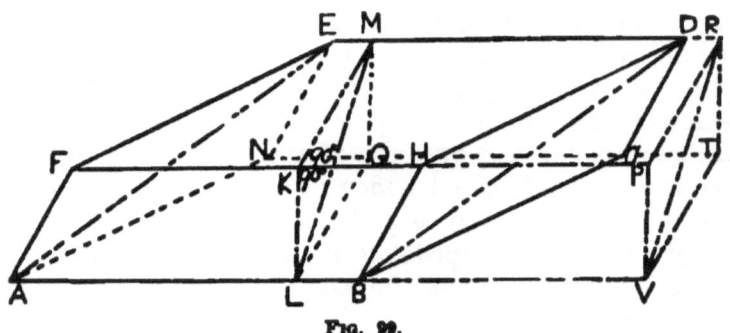

Fig. 99.

into oblique triangular prisms by a plane through the diagonally opposite edges A B and E D. Imagine also, that, before the pieces are separated, the parallelopiped is cut along the lines M K and K L, which have been drawn at right angles to E H. Transfer the portion cut off at the left to the right, forming the parallelopiped L V T R M K P. At the same time the triangular prisms B H D E F A and B C D E A N will be changed into the prisms V P R M K L, and V T R M L Q.

If possible, follow the directions just given with a block of wood or a piece of potato in the proper shape.

How many faces of the new parallelopiped will be rectangles? If the new parallelopiped is imagined to stand on the base V T R P, it will be a **right parallelopiped**, because its edges, K P, M R, etc., are at right angles to the two faces. Illustrate this by showing that your block, if placed on the base V T R P, or M K L Q, will stand upright, but that if it is placed on any other base it will lean over. The volume of the right parallelopiped, by 639, will be the product of the base V T R P by the altitude K P. The triangular prisms V P R M K L, and V T R M L Q, are right prisms. (Why?) The volume of one of them is, by 642, the product of the base V P R and the altitude K P, or just one half of the volume of the right parallelopiped. Hence a right parallelopiped, as well as a rectangular parallelopiped (640), can be cut into two parts which have the same volume by a plane through two diagonally opposite edges. But the right parallelopiped of Fig. 99 occupies the same space as the original oblique parallelopiped; and the right triangular prisms occupy the same space as the oblique triangular prisms, so that **any parallelopiped—whether rectangular, right, or oblique—can be cut into two parts which have the same volume by a plane through two diagonally opposite edges.**

644. Test the principle just stated in 643 by weighing the two pieces into which parallelopipeds of wood having the required shapes are cut by planes containing two diagonally opposite edges.

645. The volume of the oblique parallelopiped A B C D E F H (Fig. 99) is, by 639, the product of its base B C D H by its altitude, the perpendicular distance between its two bases; the volume of the oblique prism B D H E F A, which is one half of the parallelopiped by 643, is the product of one half of B C D H, or B D H, by its altitude, that is, it is the product of its base B D H by its altitude, so that **the volume of any triangular prism is the product of its base by its altitude.**

646. In how many different ways could you cut an oblique parallelopiped into triangular prisms by planes containing two diagonally opposite edges? Would there be any difference in size in the various prisms? in shape?

647. Make three lines, a, b, and c; form a right parallelopiped equivalent to the product of the lines $a \times b \times c$; also form an oblique parallelopiped equivalent to $a \times b \times c$; form an oblique triangular prism equivalent to $a \times b \times c$.

648. A tin box, in form a right triangular prism, has as its bases two right-angled isosceles triangles with the equal sides 8 cm. long; the box is 15 cm. tall. How many ccm. of water will the box hold? How many sq. cm. of tin will be needed for the box and its cover?

649. The inside dimensions of a cistern shaped like a rectangular parallelopiped are 10 ft. × 6 ft. × 8 ft.; how many gallons of water will the cistern hold, if it takes $7\frac{1}{2}$ gallons to fill a cubic foot of space?

650. A right prism 10 in. tall has a regular hexagon for a base; if each edge of the base is 6 in., how much space does the prism occupy?

651. A water-pail in the form of a cylinder is 12 in. tall and 9 in. in diameter; can you tell how many gallons it will hold if it takes 231 cu. in. to hold a gallon? (The volume of a cylinder is found on the same principle on which the volume of a prism is found.)

652. Into a tumbler partly filled with water put a piece of lead or a block of wood of irregular shape, and note carefully how much the water rises when the object is completely under water. Estimate the volume of the immersed solid after taking the proper measurements. Use a tumbler that has the form of a cylinder.

653. Sink in a tumbler of water a triangular prism, and note the height to which the water rises; next, sink a triangular pyramid with a base and an altitude equal to those respectively of the prism; how much more space does the prism occupy than the pyramid does? Try this experiment several times before giving a final answer, and, if possible, use several pairs of prisms and pyramids, making each pair according to the directions above, but varying the sizes of the blocks used.

654. Form two pyramids of the same height on triangular bases that have the same area but very different forms. Sink these pyramids in water, and note whether one occupies more space than the other.

655. Form several triangular pyramids of the same altitude and with equivalent (not equal) bases, and form one prism with an altitude equal to that of the pyramids, and with a base equivalent to one of the bases of the pyramid. See how many pyramids it takes to displace as much water as the prism displaces.

656. Sum up the results of the experiments in 653–655 by writing a principle that will enable you to obtain the volume of a triangular pyramid.

657. Can you extend the principle of 656 so that it will include any pyramid and also a cone?

658. In dealing with triangles you saw that you could move the vertex of a triangle along a line parallel to the base without altering the area of the triangle. Do your experiments in 653–655 point to any similar principle about triangular pyramids? What is the principle?

In moving the vertex of a triangle without changing the area you were obliged to move it either symparallel with the base or antiparallel to it. But in moving the vertex of a triangular pyramid, you can move it in a countless number of directions without changing the volume of the pyramid; for a line will be parallel to the triangular base of the pyramid if it is parallel to any one of the countless lines that can be imagined in the triangular base. To get a clear idea of this, place a triangular pyramid on the table; you will at once see that there are thousands of directions in which you could move the vertex without changing the volume, keeping it in a line parallel to the base or the table. You can therefore mould triangular pyramids with even more freedom than you can mould triangles.

659. By moulding triangles you can change a polygon into an equivalent triangle. In the same way, by moulding triangular pyramids you can change a pyramid which has a polygon as a base into a triangular pyramid. For an example see Fig. 100, where the pyramid V—A B C D has a quadrilateral as a base.

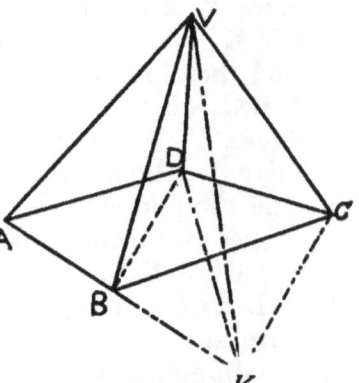

FIG. 100.

The plane V B D cuts the pyramid into two triangular

pyramids, V—A B D and V—B C D. The latter may be imagined to stand on the base B D V, with vertex at C. Move C along the line C K, which is parallel to D B, and therefore to V B D. (See 658.) The pyramid C—B D V is thus moulded into the equivalent pyramid K—B D V, which joins with V—A B D to form the triangular pyramid V—A K D; equivalent to the original pyramid.

660. Draw three different pyramids with quadrilaterals for bases, and mould each pyramid into its equivalent triangular pyramid.

661. Can you mould two triangular pyramids of the same height, but with different bases, into one triangular pyramid?

662. Review the method by which you lowered the vertex of a triangle to any desired point without changing the surface of the triangle, and then study the construction which follows, by which the vertex of a pyramid is lowered without any change in the volume of the pyramid. The original pyramid is V—A B C, Fig. 101, and the point to which the vertex is to be lowered is P.

By a plane, P B C (a guide plane corresponding to the guide line of Exercise 419), a pyramid, V—P B C, is cut off; the vertex V is then moved to K along V K, which is parallel to P B, and therefore parallel to the base P B C, which contains P B; the edges P V, B V, C V, take the respective positions, P K, B K, C K, and the pyramid V—A B C is changed into the equivalent pyramid P—A K C.

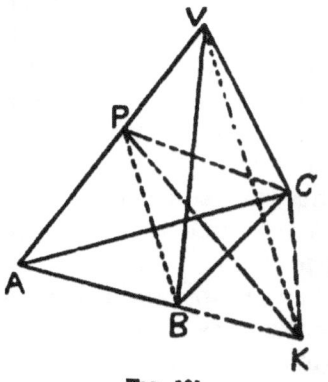

Fig. 101.

663. Draw two triangular pyramids with bases on the same plane, but with different altitudes. Lower the taller pyramid to the level of the shorter one. Then find one triangular pyramid equivalent to their sum.

664. Can you raise the vertex of a triangular pyramid to any desired level without changing the volume of the pyramid?

665. How much larger is a triangular prism than a triangular pyramid which stands on the same base and has the same altitude as the prism? (See 656.)

666. If you wish to find a triangular pyramid with a volume equal to that of a given triangular prism, how tall must you make the pyramid, if you make its base equal to the base of the prism? How large a base must it have, if you keep the height the same as that of the prism?

667. A wine-glass shaped like a cone and a tumbler shaped like a cylinder are of the same height and have the same width at the top; how much more water will the tumbler hold than the wine-glass?

668. How much more than the wine-glass would the tumbler of 667 hold, if it were twice as tall as the wine-glass, other conditions remaining the same?

669. In 432 and the following articles you saw that a circle could be moulded into a triangle with the radius as altitude, if only a line could be found for the base equal in length to the circumference, and by an experiment you found how to get the circumference from the radius with sufficient accuracy for practical purposes. In a similar way a sphere can be moulded into a cone with an altitude equal to the radius, if it is possible to find a base equivalent to the surface of the sphere. The following experiment will fix in your minds the facts that have been discovered about the surface of a sphere by more advanced principles of Geometry than you have yet taken up. Take a ball of wood or of rubber (a baseball would answer very well) and have a tin cylindrical box made, into which the ball will just fit. When the cover is on the box it should touch the ball. Sink first the ball and then the closed box in a small cylindrical jar of water, and notice how much higher the water rises in one case than in the other. You will find that the ball makes the water rise only two-thirds

as far as the box does; so that the ball occupies only two-thirds as much space as the cylinder which just surrounds it. Now imagine the ball moulded into a cone, with the radius of the ball as its altitude, and with the surface of the ball flattened into a circle for its base. The cone will be only one-half as tall as the box, for the box has the diameter of the ball for its altitude, and the cone has the radius of the ball for its altitude. Since the cone is one-half as tall as the box, to occupy as much space as the box does it would need a base six times as large as the base of the box (see 668); but since it occupies only two-thirds as much space as the box does, its base need be only two-thirds of six times as large as the base of the box, or four times the base of the box. This experiment illustrates the truth, which can be proved by the principles of Geometry, that **the surface of a sphere is four times the base of a cylinder in which the sphere is inscribed.**

670. If you should cut the ball in halves, and then replace the halves in the box, the flat side of one half would exactly fit the bottom of the box, and the flat side of the other half would fit the top of the box, while the curved portions would just touch each other in the centre of the box. What portion of the box is empty?

If you cut an orange in halves, the two circles forming the plane sides of the pieces contain as much surface as the peel on one of the pieces. Why?

671. If you should cut the box used in 669 and 670 along a line perpendicular to the two bases, and flatten the lateral surface into a plane surface, you would form a rectangle with the circumference of the original base as a base, and with an altitude equal to twice the radius of the base. If you should also mould one of the bases into a triangle, the triangle would have the circumference of the original base as a base, and an altitude equal to the radius. Since the rectangle has the same base as the triangle, but twice its altitude, the surface of the rectangle is four times that of the triangle. Make this clear by drawing a rectangle and

a triangle which fulfil the above conditions, and by changing the triangle into a rectangle. Since the lateral surface of the cylindrical box is four times that of its base, it must be equivalent to that of the ball. Hence another way of describing the surface of a sphere : **The surface of a sphere is the same as the lateral surface of a cylinder in which the sphere is inscribed.**

672. Compare the entire surface of a cylinder with that of a sphere which is inscribed in the cylinder.

673. Can you draw a circle that shall have as much surface as a given sphere has ? A sphere is given if its radius is known.

674. Can you draw a cone that shall have a base whose radius is the radius of a given sphere, and that at the same time shall equal the sphere in volume ?

675. Find by arithmetic the volume of each of the following spheres, the radii of which are 7 in., 6 in., 3 in., and 14 in., respectively. Compare the first answer with the last answer, and the second answer with the third answer.

676. Assuming that the earth is a sphere with a radius of 4,000 miles, how many square miles are there on the earth's surface ?

677. As a result of your study of this section write and record a rule for finding the volume of each of the following solids : a parallelopiped ; a prism ; a pyramid ; a cone ; a cylinder ; and a sphere.

678. Draw three lines, a, b, and c, and form a parallelopiped, a prism, a triangular pyramid, and a quadrangular pyramid, equivalent to the product of these lines. Can you draw a cone, a cylinder, and a sphere that shall be equivalent to the product of these lines ?

INDEX

ALTITUDE, 242.
Angle, 76; Right, Acute, Obtuse, Oblique, Straight Angles, 77; Exterior, Interior, Alternate Exterior or Interior, Corresponding Angles, 121; Inscribed Angle, 569.
Antiparallel, 113.
Arc, 566.
Area, 371.
Axiom, 61.

CHORD, 569.
Circumference, 32.
Contact, Point of, 588.
Corollary, 213.

DIAGONAL, 297.
Dimensions, 4.
Dividing Tool, 355.
Drawing to Scale, 183.

EDGES, 12.
Ellipse, 38.
Equilateral, 219.
Equivalent, 373.

FACES, 12.

GEOMETRICAL SOLIDS, 4.

HEXAGON, 404.
Hypotenuse, 272.

INSCRIBE, 312.
Inscribed Angle, 569.
Interior Tangent, 624.
Isosceles, 210.

LINES, 12, 26, and 27.

MEDIAN OF A TRAPEZOID, 333.

PARALLEL, 113.
Parallelogram, 148.
Parallelopiped, 627, 632, 643.
Pentagon, 97
Perimeter, 379.
Perpendicular, 224
Points, 17.
Polygon, 428.
Prism, 640.
Product of Two Lines, 368; of Three Lines, 629.
Proof, 150.
Proportional Dividers, 351.
Protractor, 83.
Pythagorean Principle, 463.

QUADRILATERAL, 100.

RADIUS, 33.
Rectangle, 290.
Rectangular Parallelopiped, 627.
Rhomboid, 290.
Rhombus, 290.

138 INDEX

Right Angle, 77; Right Prism, 640; Right Parallelopiped, 643.
Root, 508.

SCALENE, 324.
Scholium, 121.
Segment, 582.
Similar, 349.
Solids, 3, 4, 58.
Square, 290.
Square Root, 508.

Stand on, 574.
Straight Angle, 79.
Supplement, 119.
Surface, 12, 47.
Symparallel, 113.

TANGENT, 588, 624.
Trapezium, 281.
Trapezoid, 281.
Triangles, 105, 210, 324.

VERTEX, 87.

SIGNS USED IN ADDITION TO THE ORDINARY SIGNS OF ARITHMETIC.

=, is equal to.
≠, is not equal to.
<, is smaller than.
>, is larger than.
∠, ∡, angle, angles.
△, ▲, triangle, triangles.
∥, is parallel to.
↑↑, is symparallel with.
↑↓, is antiparallel to.
⊥, is perpendicular to.

⌦, is equivalent to.
∴, therefore.
▭, rectangle.
∠ ⌐, parallelogram, or rhomboid.
○, ⊙, circumference, circumferences.
⊙, ⊖, circle, circles.
⌒, arc.
√, square root.

SUMMARY OF
FACTS, DEFINITIONS, AND PRINCIPLES

SUMMARY OF FACTS, DEFINITIONS, AND PRINCIPLES

NOTE.—Each pupil should enter in this Summary the important principles and definitions, when he learns them, expressing them in the language that seems most natural to himself, provided that his words accurately express the ideas intended.

An index should be made, at the time of the entries, on the last blank page.

Each teacher will have his own opinion of the principles that he wishes his particular class to become familiar with; so that the entries here ought to vary with different teachers, and also with different classes under the same teacher.

To assist those who have no wish to make a selection for themselves, it is suggested that entries ought to be made, in general, after the study of exercises 20, 21, 22, 23, 26, 27, 32, 47, 56, 57, 58, 61, 76, 77, 79, 113, 114, 115, 118, 120, 121, 144, 149, 150, 156, 160, 161, 166, 178, 192, 193, 205, 211, 212, 213, 214, 249, 265, 272, 278, 297, 309, 318, 319, 322, 325, 333, 371, 385, 393, 463, 537, 566, 569, 571, 581, 591, 639, 642, 643, 656, 658, 669, 671, and 677.

www.ingramcontent.com/pod-product-compliance
Lightning Source LLC
Chambersburg PA
CBHW030344170426
43202CB00010B/1233